PROBLEMS AND SOLUTIONS IN REAL ANALYSIS

Series on Number Theory and Its Applications   ISSN 1793-3161

Series Editor: Shigeru Kanemitsu *(Kinki University, Japan)*

Editorial Board Members:
V. N. Chubarikov *(Moscow State University, Russian Federation)*
Christopher Deninger *(Universität Münster, Germany)*
Chaohua Jia *(Chinese Academy of Sciences, PR China)*
Jianya Liu *(Shangdong University, PR China)*
H. Niederreiter *(National University of Singapore, Singapore)*
M. Waldschmidt *(Université Pierre et Marie Curie, France)*

Advisory Board:
K. Ramachandra *(Tata Institute of Fundamental Research, India (retired))*
A. Schinzel *(Polish Academy of Sciences, Poland)*

Series on Number Theory and Its Applications Vol. 4

PROBLEMS AND SOLUTIONS IN REAL ANALYSIS

Masayoshi Hata

Kyoto University, Japan

World Scientific

NEW JERSEY • LONDON • SINGAPORE • BEIJING • SHANGHAI • HONG KONG • TAIPEI • CHENNAI

Published by

World Scientific Publishing Co. Pte. Ltd.

5 Toh Tuck Link, Singapore 596224

USA office: 27 Warren Street, Suite 401-402, Hackensack, NJ 07601

UK office: 57 Shelton Street, Covent Garden, London WC2H 9HE

British Library Cataloguing-in-Publication Data
A catalogue record for this book is available from the British Library.

Series on Number Theory and Its Applications — Vol. 4
PROBLEMS AND SOLUTIONS IN REAL ANALYSIS

Copyright © 2007 by World Scientific Publishing Co. Pte. Ltd.

ISBN-13 978-981-277-601-3
ISBN-10 981-277-601-X
ISBN-13 978-981-277-949-6 (pbk)
ISBN-10 981-277-949-3 (pbk)

Printed in Singapore.

Preface

Rome was not built in a day...

There is no shortcut to good scholarship. To learn mathematics you are to solve many 'good' problems without haste. Mathematics is not only for persons of talent. Tackling difficult problems is like challenging yourself. Even if you do not make a success in solving a problem, you may set some new knowledge or technique you lacked.

This book contains more than one hundred and fifty mathematical problems and their detailed solutions related mainly to Real Analysis. Many problems are selected carefully both for students who are presently learning or those who have just finished their courses in Calculus and Linear Algebra, or for any person who wants to review and improve his or her skill in Real Analysis and, moreover, to make a step forward, for example, to Complex Analysis, Fourier Analysis, or Lebesgue Integration, etc. The solutions to all problems are supplied in detail, which should compete well with the famous books written by Pólya and Szegö more than thirty-five years ago.

Some problems are taken from Analytic Number Theory; for example, the uniform distribution (Chapter 12) and the prime number theorem (Chapter 17). The latter is treated in a slightly different way. They may be useful for an introduction to Analytic Number Theory.

Nevertheless the reader should notice that all solutions are not short and elegant. It may always be possible for the reader to find better and more elementary solutions. The problems are merely numbered for convenience' sake and so the reader should grapple with them using any tools, which makes a difference from the usual exercises in Calculus. One may use integration for problems on series, for example. The author must confess that there are some problems expressed in an elementary way, whose simple and elementary proof could not be found by the author. The reason why he dared to include such problems and the solutions beyond the limits of Calculus is leaving to urge the reader to find better ones.

The author wishes to take this opportunity to thank Professor S. Kanemitsu for invaluable help in the preparation of the manuscript.

Enjoy mathematics with a pen !

M. Hata

Some remarks on the definitions and notations are listed below.

- Let $f(x)$ be a real-valued function defined on an open interval (a, b) and let c be any point in $[a, b)$. We write $f(c+)$ or

$$\lim_{x \to c+} f(x)$$

  to denote the right-hand limit if it exists. Note that some books use the notation "$x \to c+0$" instead of "$x \to c+$". This is the limit of $f(x)$ as x approaches to the point c satisfying $x > c$. The left-hand limit $f(c-)$ can also be defined similarly for $c \in (a, b]$ if it exists.

- The right-hand derivative of $f(x)$ at c is denoted by $f'_+(x)$ if it exists. We also define the left-hand derivative $f'_-(x)$ if it exists.

- Given two sequences $\{a_n\}$ and $\{b_n\}$ with $b_n \geq 0$ for all $n \geq 1$, we write

$$a_n = O(b_n)$$

  if $|a_n| \leq Cb_n$ holds for all $n \geq 1$ with some positive constant C. In particular, $a_n = O(1)$ is nothing but the boundedness of the sequence $\{a_n\}$. This notation of Landau is a convenient way of expressing inequalities. Note that the symbol $O(b_n)$ is *not* a specific sequence; thus one can write $O(1) + O(1) = O(1)$, for example. Usually we use these notations to show the asymptotic behavior of $\{a_n\}$. The majorant sequence $\{b_n\}$ may be chosen among standard positive sequences such as n^α, $(\log n)^\beta$, $e^{\delta n}$, etc., where

$$e = \lim_{n \to \infty} \left(1 + \frac{1}{n}\right)^n = \sum_{n=0}^{\infty} \frac{1}{n!}$$

  is the base to the natural logarithm.

- If $b_n > 0$ for all sufficiently large n and a_n/b_n converges to 0 as $n \to \infty$, then we write

$$a_n = o(b_n).$$

  In particular, $a_n = o(1)$ means simply that a_n converges to 0 as $n \to \infty$.

- If the limit of a_n/b_n exists and equals to 1, we write

$$a_n \sim b_n$$

  which gives clearly an equivalence relation. In this case we say that $\{a_n\}$ and $\{b_n\}$ are asymptotically equivalent or that $\{a_n\}$ is asymptotic to $\{b_n\}$. $\{b_n\}$ is also said to be the principal part of $\{a_n\}$. Note that $a_n \sim b_n$ if and only if

$$a_n = b_n + o(b_n).$$

- Landau's notations can also be applied to express the asymptotic behavior of a given function $f(x)$ as $x \to c$. If $|f(x)| \le Cg(x)$ holds on a sufficiently small neighborhood of the point c for some non-negative function $g(x)$ and positive constant C, we write

$$f(x) = O(g(x))$$

as $x \to c$. We can also define $f(x) = o(g(x))$ and $f(x) \sim g(x)$ in the same manner as with sequences, even when $x \to \infty$ or $x \to -\infty$.

- The set of all continuous functions $f(x)$ defined on an interval I possessing the continuous n times derivative $f^{(n)}(x)$ is denoted by $C^n(I)$. If the interval I contains an end point, the derivative at this point may be regarded as the one-sided derivative. In particular, the set of all continuous functions defined on I is denoted by $C(I)$.

- The sign function $\operatorname{sgn}(x)$ is defined by

$$\operatorname{sgn}(x) = \begin{cases} 1 & \text{for} \quad x > 0, \\ 0 & \text{for} \quad x = 0, \\ -1 & \text{for} \quad x < 0. \end{cases}$$

- We will reasonably suppress or abbreviate parentheses used in some cases. For example, we write $\sin n\theta$ and $\sin^2 x$ instead of $\sin(n\theta)$ and $(\sin x)^2$ respectively.

Contents

Chapter 1

Sequences and Limits

- Let $\{a_n\}$ be a sequence of real or complex numbers. A necessary and sufficient condition for the sequence to converge is that for any $\epsilon > 0$ there exists an integer $N > 0$ such that

$$|a_p - a_q| < \epsilon$$

holds for all integers p and q greater than N. This is called the *Cauchy criterion*.

- Any monotone bounded sequence is convergent.

- For any sequence $\{a_n\}$ the *inferior limit* and the *superior limit* are defined by the limits of monotone sequences

$$\liminf_{n \to \infty} a_n = \lim_{n \to \infty} \inf \{a_n, a_{n+1}, \ldots\}$$

and

$$\limsup_{n \to \infty} a_n = \lim_{n \to \infty} \sup \{a_n, a_{n+1}, \ldots\}$$

respectively. Note that the inferior and superior limits always exist if we adopt $\pm\infty$ as limits.

- A bounded sequence $\{a_n\}$ converges if and only if the inferior limit coincides with the superior limit.

---

2   *Problems and Solutions in Real Analysis*

PROBLEM 1.1 ————————————————————————————

Prove that $n \sin(2n! e\pi)$ converges to 2π as $n \to \infty$.

PROBLEM 1.2 ————————————————————————————

Prove that the sequence

$$\left(\frac{1}{n}\right)^n + \left(\frac{2}{n}\right)^n + \cdots + \left(\frac{n}{n}\right)^n$$

converges to $e/(e-1)$ as $n \to \infty$.

PROBLEM 1.3 ————————————————————————————

Prove that the sequence

$$e^{n/4} n^{-(n+1)/2} (1^1 \cdot 2^2 \cdots n^n)^{1/n}$$

converges to 1 as $n \to \infty$.

This was proposed by Cesàro (1888) and solved by Pólya (1911).

PROBLEM 1.4 ————————————————————————————

Suppose that a_n and b_n converge to α and β as $n \to \infty$ respectively. Show that the sequence

$$\frac{a_0 b_n + a_1 b_{n-1} + \cdots + a_n b_0}{n}$$

converges to $\alpha\beta$ as $n \to \infty$.

PROBLEM 1.5 ————————————————————————————

Suppose that $\{a_n\}_{n\geq 0}$ is a non-negative sequence satisfying

$$a_{m+n} \leq a_m + a_n + C$$

for all positive integers m, n and some non-negative constant C. Show that a_n/n converges as $n \to \infty$.

This is essentially due to Fekete (1923). In various places we encounter this useful lemma in deducing the existence of limits.

─ **PROBLEM 1.6** ─────────────────────────────────

For any positive sequence $\{a_n\}_{n \geq 1}$ show that

$$\left(\frac{a_1 + a_{n+1}}{a_n} \right)^n > e$$

for infinitely many n's, where e is base of the natural logarithm. Prove moreover that the constant e on the right-hand side cannot in general be replaced by any larger number.

─ **PROBLEM 1.7** ─────────────────────────────────

For any $0 < \theta < \pi$ and any positive integer n show the inequality

$$\sin \theta + \frac{\sin 2\theta}{2} + \cdots + \frac{\sin n\theta}{n} > 0.$$

This was conjectured by Fejér and proved by Jackson (1911) and by Gronwall (1912) independently. Landau (1934) gave a shorter (maybe the shortest) elegant proof. See also **PROBLEM 5.9**. Note that

$$\sum_{n=1}^{\infty} \frac{\sin n\theta}{n} = \frac{\pi - \theta}{2}$$

for $0 < \theta < 2\pi$, which is shown in **SOLUTION 7.10**.

─ **PROBLEM 1.8** ─────────────────────────────────

For any real number θ and any positive integer n show the inequality

$$\frac{\cos \theta}{2} + \frac{\cos 2\theta}{3} + \cdots + \frac{\cos n\theta}{n + 1} \geq -\frac{1}{2}.$$

This was shown by Rogosinski and Szegö (1928). Verblunsky (1945) gave another proof. Koumandos (2001) obtained the lower bound $-41/96$ for $n \geq 2$.
Note that

$$\sum_{n=0}^{\infty} \frac{\cos n\theta}{n + 1} = \frac{\pi - \theta}{2} \sin \theta - \cos \theta \log \left(2 \sin \frac{\theta}{2} \right)$$

for $0 < \theta < 2\pi$.
For the simpler cosine sum Young (1912) showed that

$$\sum_{n=1}^{m} \frac{\cos n\theta}{n} > -1$$

for any θ and positive integer $m \geq 2$. Brown and Koumandos (1997) improved this by replacing -1 by $-5/6$.

PROBLEM 1.9

Given a positive sequence $\{a_n\}_{n \geq 0}$ satisfying $\sqrt{a_1} \geq \sqrt{a_0} + 1$ and

$$\left| a_{n+1} - \frac{a_n^2}{a_{n-1}} \right| \leq 1$$

for any positive integer n, show that

$$\frac{a_{n+1}}{a_n}$$

converges as $n \to \infty$. Show moreover that $a_n \theta^{-n}$ converges as $n \to \infty$, where θ is the limit of the sequence a_{n+1}/a_n.

This is due to Boyd (1969).

PROBLEM 1.10

Let E be any bounded closed set in the complex plane containing an infinite number of points, and let M_n be the maximum of $|V(x_1, ..., x_n)|$ as the points $x_1, ..., x_n$ run through the set E, where

$$V(x_1, ..., x_n) = \prod_{1 \leq i < j \leq n} (x_i - x_j)$$

is the Vandermonde determinant. Show that $M_n^{2/(n(n-1))}$ converges as $n \to \infty$.

This is due to Fekete (1923) and the limit

$$\tau(E) = \lim_{n \to \infty} M_n^{2/(n(n-1))}$$

is called the *transfinite diameter* of E. See **PROBLEM 15.9**.

Solutions for Chapter 1

SOLUTION 1.1

Let r_n and ϵ_n be the integral and fractional parts of the number $n!e$ respectively. Using the expansion

$$e = 1 + \frac{1}{1!} + \frac{1}{2!} + \cdots + \frac{1}{n!} + \cdots,$$

we have

$$\begin{cases} r_n = n!\left(1 + \frac{1}{1!} + \frac{1}{2!} + \cdots + \frac{1}{n!}\right) \\ \epsilon_n = \frac{1}{n+1} + \frac{1}{(n+1)(n+2)} + \cdots, \end{cases}$$

since

$$\frac{1}{n+1} < \epsilon_n < \frac{1}{n+1} + \frac{1}{(n+1)^2} + \cdots = \frac{1}{n}.$$

Thus $\sin(2n!e\pi) = \sin(2\pi\epsilon_n)$. Note that this implies the irrationality of e.

Since $n\epsilon_n$ converges to 1 as $n \to \infty$, we have

$$\lim_{n\to\infty} n\sin(2\pi\epsilon_n) = \lim_{n\to\infty} \frac{\sin(2\pi\epsilon_n)}{\epsilon_n} = 2\pi.$$

Hence $n\sin(2n!e\pi)$ converges to 2π as $n \to \infty$. $\qquad\square$

REMARK. More precisely one gets

$$\epsilon_n = \frac{1}{n} - \frac{1}{n^3} + O\left(\frac{1}{n^4}\right);$$

hence we have

$$n\sin(2n!e\pi) = 2\pi n\epsilon_n + \frac{4\pi^3}{3}n\epsilon_n^3 + O(n\epsilon_n^5)$$

$$= 2\pi + \frac{2\pi(2\pi^2 - 3)}{3n^2} + O\left(\frac{1}{n^3}\right)$$

as $n \to \infty$.

SOLUTION 1.2

Let $\{d_n\}$ be any monotone increasing sequence of positive integers diverging to ∞ and satisfying $d_n < n$ for $n > 1$. We divide the sum into two parts as follows.

$$\begin{cases} a_n = \left(\dfrac{1}{n}\right)^n + \left(\dfrac{2}{n}\right)^n + \cdots + \left(\dfrac{n-1-d_n}{n}\right)^n, \\[2mm] b_n = \left(\dfrac{n-d_n}{n}\right)^n + \cdots + \left(\dfrac{n}{n}\right)^n. \end{cases}$$

First the sum a_n is roughly estimated above by

$$\frac{1}{n^n} \int_0^{n-d_n} x^n \, dx = \frac{(n-d_n)^{n+1}}{(n+1)n^n} < \left(1 - \frac{d_n}{n}\right)^n.$$

Now using the inequality $\log(1-x) + x < 0$ valid for $0 < x < 1$ we obtain

$$0 < a_n < e^{n\log(1-d_n/n)} < e^{-d_n},$$

which converges to 0 as $n \to \infty$.

Next by using Taylor's formula for $\log(1-x)$ we can take a positive constant c_1 such that the inequality

$$|\log(1-x) + x| \le c_1 x^2$$

holds for any $|x| \le 1/2$. Thus for any integer n satisfying $d_n/n \le 1/2$ we get

$$\left| n \log\left(1 - \frac{k}{n}\right) + k \right| \le \frac{c_1 k^2}{n} \le \frac{c_1 d_n^2}{n}$$

for $0 \le k \le d_n$. Suppose further that d_n^2/n converges to 0 as $n \to \infty$. For example $d_n = [n^{1/3}]$ satisfies all the conditions imposed above. Next take a positive constant c_2 satisfying

$$|e^x - 1| \le c_2 |x|$$

for any $|x| \le 1$. Since $c_1 d_n^2/n \le 1$ for all sufficiently large n, we have

$$\left| e^k \left(1 - \frac{k}{n}\right)^n - 1 \right| = \left| e^{n\log(1-k/n)+k} - 1 \right|$$

$$\le \frac{c_1 c_2 d_n^2}{n}.$$

Dividing both sides by e^k and summing from $k = 0$ to d_n, we get

$$\sum_{k=0}^{d_n} \left| \left(1 - \frac{k}{n}\right)^n - e^{-k} \right| \leq \frac{c_1 c_2 d_n^2}{n} \sum_{k=0}^{d_n} e^{-k}.$$

Hence

$$\left| b_n - \sum_{k=0}^{d_n} e^{-k} \right| < \frac{e c_1 c_2 d_n^2}{(e-1)n},$$

which implies

$$\left| b_n - \frac{e}{e-1} \right| \leq \frac{e}{e-1} \left(\frac{c_1 c_2 d_n^2}{n} + e^{-d_n} \right).$$

Therefore $a_n + b_n$ converges to $e/(e-1)$ as $n \to \infty$.   □

SOLUTION 1.3

One can easily verify that the function $f(x) = x \log x$ satisfies all the conditions stated in **PROBLEM 5.7**. Therefore the logarithm of the given sequence converges to

$$\frac{f(1) - f(0+)}{2} = 0,$$

hence the limit is 1.   □

SOLUTION 1.4

Let M be an upper bound of the two convergent sequences $|a_n|$ and $|b_n|$. For any $\epsilon > 0$ we can take a positive integer N satisfying $|a_n - \alpha| < \epsilon$ and $|b_n - \beta| < \epsilon$ for all integers n greater than N. If n is greater than N^2, then

$$|a_k b_{n-k} - \alpha\beta| \leq |(a_k - \alpha)b_{n-k} + \alpha(b_{n-k} - \beta)|$$
$$\leq (M + |\alpha|)\epsilon$$

for any integer k in the interval $[\sqrt{n}, n - \sqrt{n}]$. Therefore

$$\left| \frac{1}{n} \sum_{k=0}^{n} a_k b_{n-k} - \alpha\beta \right| \leq \frac{1}{n} \sum_{\sqrt{n} \leq k \leq n - \sqrt{n}} |a_k b_{n-k} - \alpha\beta|$$

$$+ 2\left(|\alpha\beta| + M^2\right) \frac{[\sqrt{n}] + 1}{n}$$

$$\leq (M + |\alpha|)\epsilon + 2\left(|\alpha\beta| + M^2\right) \frac{\sqrt{n} + 1}{n}.$$

We can take n so large that the last expression is less than $(M + |\alpha| + 1)\epsilon$. □

SOLUTION 1.5

For an arbitrary fixed positive integer k we put $n = qk + r$ with $0 \le r < k$. Since $a_n = a_{qk+r} \le q(a_k + C) + a_r$, we have

$$\frac{a_n}{n} \le \frac{a_k + C}{k} + \frac{a_r}{n}.$$

Taking the limit as $n \to \infty$, we get

$$\limsup_{n\to\infty} \frac{a_n}{n} \le \frac{a_k + C}{k}.$$

The sequence a_n/n is therefore bounded. Since k is arbitrary, we may conclude that

$$\limsup_{n\to\infty} \frac{a_n}{n} \le \liminf_{k\to\infty} \frac{a_k}{k},$$

which means the convergence of a_n/n. □

SOLUTION 1.6

Without loss of generality we may put $a_1 = 1$. Suppose, contrary to the conclusion, that there is an integer N satisfying

$$\left(\frac{1 + a_{n+1}}{a_n}\right)^n \le e$$

for all $n \ge N$. Put

$$s_{j,k} = \exp\left(\frac{1}{j} + \cdots + \frac{1}{k}\right)$$

for any integers $j \le k$. Since $0 < a_{n+1} \le e^{1/n}a_n - 1$, we get successively

$$\begin{cases} 0 < a_{n+1} \le s_{n,n}\, a_n - 1, \\ 0 < a_{n+2} \le s_{n,n+1}\, a_n - s_{n+1,n+1} - 1, \\ \qquad\qquad \vdots \\ 0 < a_{n+k+1} \le s_{n,n+k}\, a_n - s_{n+1,n+k} - \cdots - s_{n+k,n+k} - 1 \end{cases}$$

for any non-negative integer k. Hence it follows that

$$a_n > \frac{1}{s_{n,n}} + \frac{1}{s_{n,n+1}} + \cdots + \frac{1}{s_{n,n+k}}.$$

On the other hand, using the inequality

$$\frac{1}{s_{n,n+j}} > \exp\left(-\int_{n-1}^{n+j}\frac{dx}{x}\right) = \frac{n-1}{n+j},$$

we get

$$a_n > \sum_{j=0}^{k}\frac{n-1}{n+j},$$

which is a contradiction, since the right-hand side diverges to ∞ as $k \to \infty$.

To see that the bound e cannot be replaced by any larger number, consider the case $a_n = n \log n$ for $n \geq 2$. Then

$$\left(\frac{a_1 + (n+1)\log(n+1)}{n\log n}\right)^n = \exp\left(n\log\left(1 + \frac{1}{n} + O\left(\frac{1}{n\log n}\right)\right)\right)$$

$$= \exp\left(1 + O\left(\frac{1}{\log n}\right)\right),$$

which converges to e as $n \to \infty$.   □

SOLUTION 1.7

Denote by $s_n(\theta)$ the left-hand side of the inequality to be shown. Write θ for 2ϑ for brevity. Since

$$s_n'(\theta) = \Re\left(e^{i\theta} + e^{2i\theta} + \cdots + e^{ni\theta}\right)$$

$$= \frac{\cos(n+1)\vartheta \sin n\vartheta}{\sin\vartheta},$$

we obtain the candidates for extreme points of $s_n(\theta)$ on the interval $(0, \pi]$ by solving the equations $\cos(n+1)\vartheta = 0$ and $\sin n\vartheta = 0$, as follows:

$$\frac{\pi}{n+1}, \quad \frac{2\pi}{n}, \quad \frac{3\pi}{n+1}, \quad \frac{4\pi}{n}, \quad \cdots$$

where the last two candidates are $(n-1)\pi/(n+1)$ and π if n is even, and $(n-1)\pi/n$ and $n\pi/(n+1)$ if n is odd. In any case $s_n'(\theta)$ vanishes at least at n points in the interval $(0, \pi)$.

Since $s_n'(\theta)$ can be expressed as a polynomial in $\cos\theta$ of degree n and $\cos\theta$ maps the interval $[0, \pi]$ onto $[-1, 1]$ homeomorphically, this polynomial possesses at most n real roots in $[-1, 1]$. Therefore all these roots must be simple and give the actual extreme points of $s_n(\theta)$ except for $\theta = \pi$. Clearly $s_n(\theta)$ is positive in the right neighborhood of the origin, and the maximal and minimal points stand in line alternately from left to right. Thus $s_n(\theta)$ attains its minimal values at the points $2\ell\pi/n \in (0, \pi)$ when $n \geq 3$. In the cases $n = 1$ and $n = 2$, however, $s_n(\theta)$ has no minimal points in $(0, \pi)$.

Now we will show that $s_n(\theta)$ is positive on the interval $(0, \pi)$ by induction on n. This is clear for $n = 1$ and $n = 2$ since $s_1(\theta) = \sin\theta$ and $s_2(\theta) = (1+\cos\theta)\sin\theta$. Suppose that $s_{n-1}(\theta) > 0$ for some $n \geq 3$. Then the minimal values of $s_n(\theta)$ are certainly attained at some points $2\ell\pi/n$ in $(0, \pi)$, whose values are

$$s_n\left(\frac{2\ell\pi}{n}\right) = s_{n-1}\left(\frac{2\ell\pi}{n}\right) + \frac{\sin 2\ell\pi}{n}$$

$$= s_{n-1}\left(\frac{2\ell\pi}{n}\right) > 0.$$

Therefore $s_n(\theta) > 0$ on the interval $(0, \pi)$.   □

REMARK. Landau (1934) gave the following elegant shorter proof using mathematical induction on n. Suppose that $s_{n-1}(\theta) > 0$ on $(0, \pi)$. If s_n attains the non-positive minimum at some point, say θ^*, then $s'_n(\theta^*) = 0$ implies

$$\sin\left(n + \frac{1}{2}\right)\theta^* = \sin\frac{\theta^*}{2}$$

and hence

$$\cos\left(n + \frac{1}{2}\right)\theta^* = \pm\cos\frac{\theta^*}{2}.$$

Since

$$\sin n\theta^* = \sin\left(n + \frac{1}{2}\right)\theta^* \cos\frac{\theta^*}{2} - \cos\left(n + \frac{1}{2}\right)\theta^* \sin\frac{\theta^*}{2}$$

$$= \sin\frac{\theta^*}{2}\cos\frac{\theta^*}{2} \pm \cos\frac{\theta^*}{2}\sin\frac{\theta^*}{2},$$

being equal either to 0 or $\sin\theta^* \geq 0$ according to the sign. We are led to a contradiction.

SOLUTION 1.8

The proof is substantially based on Verblunsky (1945). Write θ for 2ϑ for brevity. Let $c_n(\vartheta)$ be the left-hand side of the inequality to be shown. It suffices to confine ourselves to the interval $[0, \pi/2]$. Clearly $c_1(\vartheta) = \cos\vartheta/2 \geq -1/2$ and

$$c_2(\vartheta) = \frac{2}{3}\cos^2\vartheta + \frac{1}{2}\cos\vartheta - \frac{1}{3} \geq -\frac{41}{96},$$

and we assume that $n \geq 3$. Note that

$$
\cos n\theta = \frac{\sin(2n+1)\vartheta - \sin(2n-1)\vartheta}{2\sin\vartheta}
$$

$$
= \frac{\sin^2(n+1)\vartheta - 2\sin^2 n\vartheta + \sin^2(n-1)\vartheta}{2\sin^2\vartheta},
$$

whose numerator is the second difference of the positive sequence $\{\sin^2 n\vartheta\}$. Using this formula we get

$$
c_n(\vartheta) = \frac{1}{2\sin^2\vartheta} \sum_{k=1}^{n} \frac{\sin^2(k+1)\vartheta - 2\sin^2 k\vartheta + \sin^2(k-1)\vartheta}{k+1},
$$

which can be written as

$$
\frac{1}{2\sin^2\vartheta}\left(-\frac{2\sin^2\vartheta}{3} + \frac{\sin^2 2\vartheta}{12} + \cdots + \frac{2\sin^2(n-1)\vartheta}{n(n^2-1)}\right.
$$
$$
\left. -\frac{(n-1)\sin^2 n\vartheta}{n(n+1)} + \frac{\sin^2(n+1)\vartheta}{n+1}\right).
$$

Hence we obtain

$$
c_n(\vartheta) \geq -\frac{1}{3} + \frac{\cos^2\vartheta}{6} + \frac{\sin^2(n+1)\vartheta - \sin^2 n\vartheta}{2(n+1)\sin^2\vartheta}
$$

$$
= -\frac{1}{6} - \frac{\sin^2\vartheta}{6} + \frac{\sin(2n+1)\vartheta}{2(n+1)\sin\vartheta}.
$$

For any ϑ satisfying $\sin(2n+1)\vartheta \geq 0$ we obviously have $c_n(\vartheta) \geq -1/3$. Moreover if ϑ belongs to the interval $(3\pi/(2n+1), \pi/2)$, then using Jordan's inequality $\sin\vartheta \geq 2\vartheta/\pi$,

$$
c_n(\vartheta) \geq -\frac{1}{3} - \frac{1}{2(n+1)\sin(3\pi/(2n+1))}
$$

$$
\geq -\frac{1}{3} - \frac{2n+1}{12(n+1)} > -\frac{1}{2}.
$$

Thus it suffices to consider the interval $[\pi/(2n+1), 2\pi/(2n+1)]$.

In general, we consider an interval of the form

$$
\left[\frac{\alpha\pi}{2n+1}, \frac{\beta\pi}{2n+1}\right].
$$

For any ϑ satisfying $\sin(2n+1)\vartheta \leq c$ on this interval it follows that

$$
c_n(\vartheta) \geq -\frac{1}{6} - \frac{\sin^2\vartheta}{6} - \frac{c}{2(n+1)\sin\vartheta}.
$$

Now the right-hand side can be written as $-1/6 - \varphi(\sin \vartheta)$, where $\varphi(x)$ is a concave function; hence, the maximum of φ is attained at an end point of that interval. By using

$$\alpha\pi \sin \vartheta \geq 7\vartheta \sin \frac{\alpha\pi}{7},$$

we get

$$\varphi\left(\sin \frac{\alpha\pi}{2n+1}\right) = \frac{1}{6} \sin^2 \frac{\alpha\pi}{2n+1} + \frac{c}{2(n+1)\sin(\alpha\pi/(2n+1))}$$

$$\leq \frac{(\alpha\pi)^2}{6(2n+1)^2} + \frac{c}{2(n+1)} \cdot \frac{2n+1}{7\sin(\alpha\pi/7)}.$$

Since $n \geq 3$, the last expression is less than

$$\frac{(\alpha\pi)^2}{294} + \frac{c}{7\sin(\alpha\pi/7)}.$$

Similarly we get an estimate for another end point.

For $\alpha = 1$ and $\beta = 4/3$ we can take $c = \sqrt{3}/2$ so that the value of φ at the corresponding end point is less than 0.319 and 0.28 respectively. Similarly for $\alpha = 4/3$ and $\beta = 2$ we can take $c = 1$ so that the value of φ is less than 0.314 and 0.318 respectively. Therefore the maximum of φ on the interval $[\pi/(2n+1), 2\pi/(2n+1)]$ is less than $1/3$, which implies that $c_n(\vartheta) > -1/2$. $\square$

SOLUTION 1.9

We first show that

$$\frac{a_{n+1}}{a_n} > 1 + \frac{1}{\sqrt{a_0}} \tag{1.1}$$

by induction on n. When $n = 0$ this holds by the assumption. Put $\alpha = 1 + 1/\sqrt{a_0}$ for brevity. Suppose that (1.1) holds for $n \leq m$. We then have $a_k > \alpha^k a_0$ for $1 \leq k \leq m+1$. Thus

$$\left|\frac{a_{m+2}}{a_{m+1}} - \frac{a_1}{a_0}\right| \leq \sum_{k=1}^{m+1} \left|\frac{a_{k+1}}{a_k} - \frac{a_k}{a_{k-1}}\right| \leq \sum_{k=1}^{m+1} \frac{1}{a_k},$$

which is less than

$$\frac{1}{a_0} \sum_{k=1}^{m+1} \alpha^{-k} < \frac{1}{a_0(\alpha-1)} = \frac{1}{\sqrt{a_0}}.$$

Therefore

$$\frac{a_{m+2}}{a_{m+1}} > \frac{a_1}{a_0} - \frac{1}{\sqrt{a_0}} > 1 + \frac{1}{\sqrt{a_0}};$$

thus (1.1) holds also for $n = m + 1$.

Let $p > q$ be any positive integers. In the same way,

$$\left| \frac{a_{p+1}}{a_p} - \frac{a_{q+1}}{a_q} \right| \leq \sum_{k=q+1}^{p} \left| \frac{a_{k+1}}{a_k} - \frac{a_k}{a_{k-1}} \right| \leq \sum_{k=q+1}^{p} \frac{1}{a_k},$$

which is less than

$$\frac{1}{a_q} \sum_{k=1}^{p-q} \frac{1}{\alpha^k} < \frac{\sqrt{a_0}}{a_q}.$$

This means that the sequence $\{a_{n+1}/a_n\}$ satisfies the Cauchy criterion since a_q diverges to ∞ as $q \to \infty$. Letting $p \to \infty$ in the above inequalities, we get

$$\left| \frac{a_{q+1}}{a_q} - \theta \right| \leq \frac{\sqrt{a_0}}{a_q}.$$

Multiplying both sides by a_q/θ^{q+1}, we have

$$\left| \frac{a_{q+1}}{\theta^{q+1}} - \frac{a_q}{\theta^q} \right| \leq \frac{\sqrt{a_0}}{\theta^{q+1}},$$

which shows that the sequence $\{a_n/\theta^n\}$ also satisfies the Cauchy criterion. □

$\boxed{\text{SOLUTION 1.10}}$

Let $\xi_1, \ldots, \xi_{n+1}$ be the points at which $|V(x_1, \ldots, x_{n+1})|$ attains its maximum M_{n+1}. Since

$$\frac{V(\xi_1, \ldots, \xi_{n+1})}{V(\xi_1, \ldots, \xi_n)} = (\xi_1 - \xi_{n+1}) \cdots (\xi_n - \xi_{n+1}),$$

we have

$$\frac{M_{n+1}}{M_n} \leq |\xi_1 - \xi_{n+1}| \cdots |\xi_n - \xi_{n+1}|.$$

Applying the same argument to each point $\xi_1, \ldots, \xi_n$, we get $n + 1$ similar inequalities whose product gives

$$\left(\frac{M_{n+1}}{M_n} \right)^{n+1} \leq \prod_{i \neq j} |\xi_i - \xi_j| = M_{n+1}^2.$$

Hence the sequence $M_n^{2/(n(n-1))}$ is monotone decreasing. $\qquad\square$

Chapter 2

Infinite Series

- An infinite series $\sum_{n=1}^{\infty} a_n$ converges if and only if for any $\epsilon > 0$ there exists an integer $N > 0$ satisfying $|a_q + \cdots + a_p| < \epsilon$ for all integers p and q greater than N.

- An infinite series $\sum_{n=1}^{\infty} a_n$ is said to *converge absolutely* if $\sum_{n=1}^{\infty} |a_n|$ converges.

- If $\sum_{n=1}^{\infty} a_n$ converges but $\sum_{n=1}^{\infty} |a_n|$ diverges, then $\sum_{n=1}^{\infty} a_n$ is said to *converge conditionally*.

- An absolutely convergent series converges to the same sum in whatever order the terms are taken.

- Any conditionally convergent series can always be rearranged to yield a series which converges to any sum prescribed whatever, or diverges to ∞ or to $-\infty$.

PROBLEM 2.1

As is well-known, the harmonic series

$$\frac{1}{1} + \frac{1}{2} + \frac{1}{3} + \cdots + \frac{1}{n} + \cdots$$

diverges to ∞. Show, however, that the convergence of the subseries removing all terms containing the digit "7" in the decimal expression of the denominator.

This is due to Kempner (1914).

PROBLEM 2.2

Given two series $\sum_{n=0}^{\infty} a_n$ and $\sum_{n=0}^{\infty} b_n$,

$$\sum_{n=0}^{\infty} (a_0 b_n + a_1 b_{n-1} + \cdots + a_n b_0)$$

is called the Cauchy product of $\sum a_n$ and $\sum b_n$.

(a) Suppose that $\sum a_n$ and $\sum b_n$ converge to α and β respectively and that the Cauchy product converges to δ. Show then that $\alpha\beta$ is equal to δ.

(b) Suppose that $\sum a_n$ converges absolutely to α and that $\sum b_n$ converges to β. Show that the Cauchy product converges to $\alpha\beta$.

(c) Suppose that $\sum a_n$ and $\sum b_n$ converge absolutely to α and β respectively. Show that the Cauchy product also converges absolutely to $\alpha\beta$.

(d) Give an example of two convergent series whose Cauchy product is divergent.

The assertion (b) is due to Mertens (1875).

PROBLEM 2.3

For any positive sequence $\{a_n\}_{n\geq 1}$ show the inequality

$$\sum_{n=1}^{\infty} (a_1 a_2 \cdots a_n)^{1/n} < e \sum_{n=1}^{\infty} a_n.$$

Prove further that the constant e on the right-hand side cannot in general be replaced by any smaller number.

Using Lagrange multipliers Carleman (1922) showed this inequality with equality sign for non-negative sequences. At least four another proofs are known. Pólya (1926) proved that the equality cannot occur unless all a_n vanishes. Knopp (1928) gave a simpler but technical proof using the arithmetic-geometric mean inequality. Carleson (1954) proved it as an application of the integral inequality

$$\int_0^{\infty} \exp\left(-\frac{f(x)}{x}\right) dx \leq e \int_0^{\infty} \exp(-f'(x)) dx.$$

See **PROBLEM 9.9** for details. Redheffer (1967) also gave another proof by introducing further parameters into the problem. See a nice survey of Duncan and McGregor (2003) for details, in which they claimed that, in many proofs, the elementary inequality

$$\left(1 + \frac{1}{n}\right)^n < e$$

holding for all positive integer n, plays a significant role.

PROBLEM 2.4

For any positive sequence $\{a_n\}_{n \geq 1}$ show the inequality

$$\left(\sum_{n=1}^{\infty} a_n\right)^4 < \pi^2 \left(\sum_{n=1}^{\infty} a_n^2\right)\left(\sum_{n=1}^{\infty} n^2 a_n^2\right).$$

Prove further that the constant π^2 on the right-hand side cannot in general be replaced by any smaller number.

This is due to Carlson (1935), who also derived the integral version:

$$\left(\int_0^{\infty} f(x)\,dx\right)^4 \leq \pi^2 \left(\int_0^{\infty} f^2(x)\,dx\right)\left(\int_0^{\infty} x^2 f^2(x)\,dx\right). \qquad (2.1)$$

Note that the equality holds for $f(x) = (1 + x^2)^{-1}$. To my surprise Carlson improved the inequality in the above problem using (2.1). For details see the remark after **SOLUTION 2.4**.

PROBLEM 2.5

Show that the convergence of $\sum_{n=1}^{\infty} n a_n$ implies that of $\sum_{n=1}^{\infty} a_n$.

PROBLEM 2.6

Show that $\dfrac{1}{n}\sum_{k=1}^{n} a_k$ converges to 0 when $\sum_{n=1}^{\infty} \dfrac{a_n}{n}$ converges.

Problems and Solutions in Real Analysis

PROBLEM 2.7

Suppose that $\{a_n\}_{n \geq 1}$ is a non-negative sequence satisfying $\sum_{n=1}^{\infty} a_n = \infty$. Then show that the series

$$\sum_{n=1}^{\infty} \frac{a_n}{(a_1 + a_2 + \cdots + a_n)^{\alpha}}$$

converges when $\alpha > 1$, and diverges when $0 < \alpha \leq 1$.

PROBLEM 2.8

Suppose that $\sum_{n=1}^{\infty} a_n b_n$ converges for any sequence $\{b_n\}$ converging. Then show that $\sum_{n=1}^{\infty} a_n$ converges absolutely.

PROBLEM 2.9

Suppose that $\sum_{n=1}^{\infty} a_n b_n$ converges for any sequence $\{b_n\}$ such that $\sum_{n=1}^{\infty} b_n^2$ converges. Then show that $\sum_{n=1}^{\infty} a_n^2$ also converges.

PROBLEM 2.10

The limit

$$\gamma = \lim_{n \to \infty} \left(1 + \frac{1}{2} + \cdots + \frac{1}{n} - \log n \right)$$
$$= 0.57721\,56649\,01532\,86060\,65120...$$

is called Euler's constant or sometimes the Euler-Mascheroni constant. Show that the following series converges to γ:

$$\frac{1}{2} - \frac{1}{3} + 2\left(\frac{1}{4} - \frac{1}{5} + \frac{1}{6} - \frac{1}{7} \right) + 3\left(\frac{1}{8} - \frac{1}{9} + \cdots - \frac{1}{15} \right) + \cdots .$$

Vacca (1910) proved this formula and stated that it is simple and has its natural place near to the Gregory-Leibniz series

$$\frac{\pi}{4} = 1 - \frac{1}{3} + \frac{1}{5} - \frac{1}{7} + \cdots$$

and Mercator's series

$$\log 2 = 1 - \frac{1}{2} + \frac{1}{3} - \frac{1}{4} + \cdots .$$

It is even not known whether γ is rational or irrational, though conjectured to be transcendental. Hence it is desirable to approximate γ by rational numbers. Hilbert mentioned that the irrationality of γ is an unsolved problem that seems unapproachable. Nowadays the numerical value of γ is computed to more than 100 million decimal places. Papanikolaou pointed out that, if γ were rational, then the denominator would have at least 242080 digits. Improbable!

PROBLEM 2.11

Making use of the formula

$$\frac{\sin(2n+1)\theta}{(2n+1)\sin\theta} = \prod_{k=1}^{n}\left(1 - \frac{\sin^2\theta}{\sin^2 k\pi/(2n+1)}\right),$$

show that

$$\frac{\sin\pi x}{\pi x} = \prod_{n=1}^{\infty}\left(1 - \frac{x^2}{n^2}\right)$$

holds for all real x.

This is due to Kortram (1996). This product representation of $\sin x$ is usually proved in Complex Analysis as an application of the canonical product of an entire function of order 1.

Solutions for Chapter 2

SOLUTION 2.1

Any integer in the interval $[10^{k-1}, 10^k)$ has k digits in its decimal expansion. Among these there exist exactly $8 \cdot 9^{k-1}$ integers which do not contain the digit '7' in their decimal expansions. Thus the sum of the subseries defined in the problem, say S, can be estimated from above as

$$S < \sum_{k=1}^{\infty} \frac{8 \cdot 9^{k-1}}{10^{k-1}} = 80.$$

$\square$

SOLUTION 2.2

(a) Since $a_n \to 0$ and $b_n \to 0$ as $n \to \infty$, both power series

$$f(x) = \sum_{n=0}^{\infty} a_n x^n \quad \text{and} \quad g(x) = \sum_{n=0}^{\infty} b_n x^n$$

converge absolutely for $|x| < 1$. Hence the product

$$f(x)g(x) = \sum_{n=0}^{\infty} (a_0 b_n + a_1 b_{n-1} + \cdots + a_n b_0) x^n$$

also converges for $|x| < 1$. It follows from Abel's continuity theorem (**PROBLEM 7.3**) that $f(x)$, $g(x)$ and $f(x)g(x)$ converge to α, β and $\alpha\beta$ as $x \to 1-$ respectively. Thus $\delta = \alpha\beta$.

(b) Let

$$M = \sum_{n=0}^{\infty} |a_n| \quad \text{and} \quad s_n = b_0 + b_1 + \cdots + b_n$$

with $|s_n| \le K$ for some constant $K > 0$. By (a) it suffices to show the convergence of the Cauchy product. To see this put

$$c_n = \sum_{k=0}^{n} (a_0 b_k + a_1 b_{k-1} + \cdots + a_k b_0)$$

$$= a_0 s_n + a_1 s_{n-1} + \cdots + a_n s_0.$$

For any $\epsilon > 0$ there exists an integer N satisfying $|s_p - s_q| < \epsilon$ and

$$|a_q| + \cdots + |a_p| < \epsilon$$

for any integers p and q with $p > q \geq N$. Then for all $p > q > 2N$ we get

$$|c_p - c_q| = \left| \sum_{k=0}^{N} a_k(s_{p-k} - s_{q-k}) + \sum_{k=N+1}^{q} a_k(s_{p-k} - s_{q-k}) + \sum_{k=q+1}^{p} a_k s_{p-k} \right|.$$

Clearly the first sum on the right-hand side is estimated above by

$$M \max_{0 \leq k \leq N} |s_{p-k} - s_{q-k}| < M\epsilon.$$

The second and the third sums are similarly estimated above by

$$2K \sum_{k=N+1}^{p} |a_k| < 2K\epsilon.$$

This implies that the sequence $\{c_n\}$ satisfies the Cauchy criterion.

(c) The Cauchy product of $\sum_{n=0}^{\infty} |a_n|$ and $\sum_{n=0}^{\infty} |b_n|$ converges by (b).

(d) For example, take

$$a_n = b_n = \frac{(-1)^n}{\sqrt{n+1}}.$$

Obviously $\sum a_n$ and $\sum b_n$ converge. However we have

$$|a_0 b_n + a_1 b_{n-1} + \cdots + a_n b_0| = \sum_{k=1}^{n+1} \frac{1}{\sqrt{k(n+2-k)}}$$

$$\geq \sum_{k=1}^{n+1} \frac{2}{k+n+2-k},$$

which shows the divergence of the Cauchy product, since the last expression is greater than 1. □

SOLUTION 2.3

The proof is based on Pólya (1926). Let $\{b_n\}$ be an arbitrary positive sequence. First we write

$$\sum_{n=1}^{m} (a_1 a_2 \cdots a_n)^{1/n} = \sum_{n=1}^{m} \left(\frac{a_1 b_1 a_2 b_2 \cdots a_n b_n}{b_1 b_2 \cdots b_n} \right)^{1/n}.$$

Using the arithmetic-geometric mean inequality on the right-hand side, the above sum is less than or equal to

$$\sum_{n=1}^{m} \frac{1}{(b_1 b_2 \cdots b_n)^{1/n}} \cdot \frac{a_1 b_1 + a_2 b_2 + \cdots + a_n b_n}{n}$$

$$= \sum_{k=1}^{m} a_k b_k \sum_{n=k}^{m} \frac{1}{n (b_1 b_2 \cdots b_n)^{1/n}}.$$

We now take

$$b_n = n \left(1 + \frac{1}{n} \right)^n$$

so that $(b_1 b_2 \cdots b_n)^{1/n} = n + 1$. Therefore we have

$$\sum_{k=1}^{m} a_k b_k \sum_{n=k}^{m} \frac{1}{n(n+1)} = \sum_{k=1}^{m} a_k b_k \left(\frac{1}{k} - \frac{1}{m+1} \right)$$

$$< \sum_{k=1}^{m} a_k \left(1 + \frac{1}{k} \right)^k$$

and this is smaller than $e \sum_{k=1}^{m} a_k$. Note that the equality does not occur in the original inequality when $m \to \infty$.

To see that e cannot be replaced by any smaller number, we take, for example,

$$a_n = \begin{cases} n^{-1} & \text{for} \quad 1 \le n \le m, \\ 2^{-n} & \text{for} \quad n > m, \end{cases}$$

where m is an integer parameter. Then it is not hard to see that

$$\sum_{n=1}^{\infty} (a_1 a_2 \cdots a_n)^{1/n} = \sum_{n=1}^{m} n!^{-1/n} + O(1)$$

$$= e \log m + O(1).$$

and

$$\sum_{n=1}^{\infty} a_n = 1 + \sum_{n=1}^{m} \frac{1}{n} = \log m + O(1),$$

which implies that the ratio of the above two sums converges to e as $m \to \infty$. $\quad \square$

SOLUTION 2.4

Put $\alpha = \sum_{n=1}^{\infty} a_n^2$ and $\beta = \sum_{n=1}^{\infty} n^2 a_n^2$ for brevity. We of course assume that β is finite. Introducing two positive parameters σ and τ we first write

$$\left(\sum_{n=1}^{\infty} a_n \right)^2 = \left(\sum_{n=1}^{\infty} \frac{a_n \sqrt{\sigma + \tau n^2}}{\sqrt{\sigma + \tau n^2}} \right)^2.$$

Using the Cauchy-Schwarz inequality, this is less than or equal to

$$\sum_{n=1}^{\infty} \frac{1}{\sigma + \tau n^2} \sum_{n=1}^{\infty} a_n^2 (\sigma + \tau n^2) = (\alpha \sigma + \beta \tau) \sum_{n=1}^{\infty} \frac{1}{\sigma + \tau n^2}.$$

Since the function $1/(\sigma + \tau x^2)$ is monotone decreasing on $[0, \infty)$, we obtain

$$\sum_{n=1}^{\infty} \frac{1}{\sigma + \tau n^2} < \int_0^{\infty} \frac{dx}{\sigma + \tau x^2} = \frac{\pi}{2\sqrt{\sigma \tau}},$$

which implies that

$$\left(\sum_{n=1}^{\infty} a_n \right)^2 < \frac{\pi}{2} \cdot \frac{\alpha \sigma + \beta \tau}{\sqrt{\sigma \tau}}.$$

The right-hand side attains its minimum $\pi \sqrt{\alpha \beta}$ at $\sigma / \tau = \beta / \alpha$.

To see that π^2 cannot be replaced by any smaller number, we take, for example,

$$a_n = \frac{\sqrt{\rho}}{\rho + n^2}$$

with a positive parameter ρ. It then follows from

$$\sqrt{\rho} \int_1^{\infty} \frac{dx}{\rho + x^2} < \sum_{n=1}^{\infty} a_n < \sqrt{\rho} \int_0^{\infty} \frac{dx}{\rho + x^2}$$

that

$$\sum_{n=1}^{\infty} a_n = \frac{\pi}{2} + O\left(\frac{1}{\sqrt{\rho}} \right)$$

as $\rho \to \infty$. Similarly we have

$$\sum_{n=1}^{\infty} a_n^2 = \frac{\pi}{4\sqrt{\rho}} + O\left(\frac{1}{\rho}\right)$$

and

$$\sum_{n=1}^{\infty} n^2 a_n^2 = \sqrt{\rho} \sum_{n=1}^{\infty} a_n - \rho \sum_{n=1}^{\infty} a_n^2 = \frac{\pi}{4}\sqrt{\rho} + O(1),$$

which imply that

$$\left(\sum_{n=1}^{\infty} a_n\right)^4 = \frac{\pi^4}{16} + O\left(\frac{1}{\sqrt{\rho}}\right),$$

and

$$\sum_{n=1}^{\infty} a_n^2 \sum_{n=1}^{\infty} n^2 a_n^2 = \frac{\pi^2}{16} + O\left(\frac{1}{\sqrt{\rho}}\right)$$

as $\rho \to \infty$. □

REMARK. Carlson (1935) obtained the following sharp inequality:

$$\left(\sum_{n=1}^{\infty} a_n\right)^4 < \pi^2 \sum_{n=1}^{\infty} a_n^2 \sum_{n=1}^{\infty} \left(n^2 - n + \frac{7}{16}\right) a_n^2. \qquad (2.2)$$

To see this consider the function

$$f_N(x) = e^{-x/2} \sum_{n=1}^{N} (-1)^{n-1} a_n L_{n-1}(x)$$

where

$$L_n(x) = \frac{e^x}{n!} \frac{d^n}{dx^n} (x^n e^{-x}) = \sum_{k=0}^{n} \binom{n}{k} \frac{(-x)^k}{k!}$$

is called the nth Laguerre polynomial, which forms a system of orthogonal polynomials over $(0, \infty)$ with weight function e^{-x}. For brevity we write

$$(f, g) = \int_0^{\infty} f(x)g(x)e^{-x} \, dx.$$

We then have

$$\int_0^{\infty} f_N(x) \, dx = 2 \sum_{n=1}^{N} a_n,$$

$$\int_0^\infty f_N^2(x)\,dx = \sum_{m,n=1}^N (-1)^{m+n} a_m a_n (\phi_{m-1}, \phi_{n-1}) = \sum_{n=1}^N a_n^2$$

and

$$\int_0^\infty x^2 f^2(x)\,dx = \sum_{m,n=1}^N (-1)^{m+n} a_m a_n (x^2 L_{m-1}, L_{n-1})$$

$$= 2 \sum_{n=3}^N (n-1)(n-2) a_n a_{n-2}$$

$$+ 8 \sum_{n=2}^N (n-1)^2 a_n a_{n-1} + 2 \sum_{n=1}^N (3n^2 - 3n + 1) a_n^2.$$

Using (2. 1) and letting $N \to \infty$ we get a certain inequality with the sign of equality. We then use the Cauchy-Schwarz inequality for each n to obtain

$$\sum_{n=3}^\infty (n-1)(n-2) a_n a_{n-2} < \sum_{n=1}^\infty \frac{(n-1)^2 + n^2}{2} a_n^2$$

and

$$\sum_{n=2}^\infty (n-1)^2 a_n a_{n-1} < \sum_{n=1}^\infty \frac{(n-1)^2 + n^2}{2} a_n^2,$$

which imply Carlson's improved inequality (2. 2). Note that both inequalities exclude the sign of equality.

The constant $7/16$ in (2. 2) may be replaced by $3/8$, since

$$\sum_{n=2}^\infty (n-1)^2 a_n a_{n-1} = \sum_{n=2}^\infty (n-1)\left(n - \frac{3}{2}\right) a_n a_{n-1} + \frac{1}{2} \sum_{n=2}^\infty (n-1) a_n a_{n-1}$$

$$< \sum_{n=1}^\infty \left(\frac{(n-1)^2 + (n-1/2)^2}{2} + \frac{1}{2} \cdot \frac{n-1+n}{2}\right) a_n^2$$

$$= \sum_{n=1}^\infty \left(n^2 - n + \frac{3}{8}\right) a_n^2.$$

SOLUTION 2. 5

Put

$$b_n = a_1 + 2a_2 + \cdots + n a_n$$

and let M be the least upper bound of $|b_n|$. For any $\epsilon > 0$ and any integers p, q with $p > q > 2M/\epsilon$ we have

$$
\left| \sum_{n=q}^{p} a_n \right| = \left| \frac{b_q - b_{q-1}}{q} + \frac{b_{q+1} - b_q}{q+1} + \cdots + \frac{b_p - b_{p-1}}{p} \right|
$$

$$
= \left| -\frac{b_{q-1}}{q} + \left(\frac{1}{q} - \frac{1}{q+1} \right) b_q + \cdots + \left(\frac{1}{p-1} - \frac{1}{p} \right) b_{p-1} + \frac{b_p}{p} \right|
$$

$$
\leq \frac{2M}{q},
$$

which is clearly less than ϵ. This is nothing but the Cauchy criterion for the convergence of the series $\sum a_n$.   □

SOLUTION 2.6

Put

$$
\frac{a_1}{1} + \frac{a_2}{2} + \cdots + \frac{a_n}{n} = \alpha + \epsilon_n \quad \text{and} \quad \sigma_n = \frac{a_1 + a_2 + \cdots + a_n}{n},
$$

where ϵ_n is a sequence converging to 0.

We now show that

$$
\sigma_n = \frac{\alpha}{n} - \frac{\epsilon_1 + \epsilon_2 + \cdots + \epsilon_{n-1}}{n} + \epsilon_n \tag{2.3}
$$

by induction on n. The case $n = 1$ is clear, since $\sigma_1 = a_1 = \alpha + \epsilon_1$. Suppose next that (2.3) holds for $n = m$. We then have

$$
\sigma_{m+1} = \frac{m}{m+1} \sigma_m + \frac{a_{m+1}}{m+1}
$$

$$
= \frac{m}{m+1} \left(\frac{\alpha}{m} - \frac{\epsilon_1 + \cdots + \epsilon_{m-1}}{m} + \epsilon_m \right) + \epsilon_{m+1} - \epsilon_m
$$

$$
= \frac{\alpha}{m+1} - \frac{\epsilon_1 + \cdots + \epsilon_m}{m+1} + \epsilon_{m+1},
$$

as required. Obviously (2.3) implies that σ_n converges to 0 as $n \to \infty$.   □

SOLUTION 2.7

Put $s_n = a_1 + a_2 + \cdots + a_n$. The convergence is clear when $\alpha > 1$, since

$$
\sum_{n=2}^{\infty} \frac{a_n}{s_n^\alpha} \leq \sum_{n=2}^{\infty} \int_{s_{n-1}}^{s_n} \frac{dx}{x^\alpha} = \int_{s_1}^{\infty} \frac{dx}{x^\alpha} < \infty.
$$

When $0 < \alpha \leq 1$, it suffices to consider the case $\alpha = 1$ since $s_n > 1$ for all sufficiently large n. Suppose first that $s_{n-1} < a_n$ for infinitely many n's. We then have $a_n/s_n > 1/2$ for infinitely many n's, which shows the divergence of the series. Consider next the case in which $s_{n-1} \geq a_n$ for all integers n greater than some integer N. Then

$$\sum_{n>N} \frac{a_n}{s_n} \geq \frac{1}{2} \sum_{n>N} \frac{a_n}{s_{n-1}} \geq \frac{1}{2} \sum_{n>N} \int_{s_{n-1}}^{s_n} \frac{dx}{x}$$

$$= \frac{1}{2} \int_{s_N}^{\infty} \frac{dx}{x} = \infty.$$

Note that the divergence of the sequence s_n is essential in the second case. □

SOLUTION 2.8

Without loss of generality we can assume that $\{a_n\}$ is a non-negative sequence, since we can take $-b_n$ instead of b_n. Suppose, on the contrary, that the series $\sum a_n$ diverges to ∞. It follows from the result of **PROBLEM 2.7** that

$$\sum_{n=1}^{\infty} \frac{a_n}{a_1 + a_2 + \cdots + a_n} = \infty,$$

contrary to the assumption since

$$b_n = \frac{1}{a_1 + a_2 + \cdots + a_n}$$

converges to 0 as $n \to \infty$. □

SOLUTION 2.9

Suppose, on the contrary, that $\sum a_n^2$ diverges to ∞. By the result of **PROBLEM 2.7** we have

$$\sum_{n=1}^{\infty} \frac{a_n^2}{(a_1^2 + a_2^2 + \cdots + a_n^2)^2} < \infty,$$

contrary to the assumption since

$$b_n = \frac{a_n}{a_1^2 + a_2^2 + \cdots + a_n^2}$$

satisfies

$$\sum_{n=1}^{\infty} a_n b_n = \sum_{n=1}^{\infty} \frac{a_n^2}{a_1^2 + a_2^2 + \cdots + a_n^2} = \infty.$$

□

SOLUTION 2.10

For brevity put

$$\sigma_n = 1 - \frac{1}{2} + \frac{1}{3} - \cdots + \frac{1}{2^n - 1} - \log 2$$

for any positive integer n. It is easily seen that

$$\sigma_1 + \sigma_2 + \cdots + \sigma_n = 1 + \frac{1}{2} + \cdots + \frac{1}{2^n - 1} - n \log 2;$$

therefore

$$\gamma = \lim_{n \to \infty} (\sigma_1 + \sigma_2 + \cdots + \sigma_n).$$

We have Vacca's formula by noticing that $\sigma_n = \tau_n + \tau_{n+1} + \cdots$, where

$$\tau_n = \frac{1}{2^n} - \frac{1}{2^n + 1} + \frac{1}{2^n + 2} - \cdots - \frac{1}{2^{n+1} - 1}.$$

$\square$

REMARK. Hardy (1912) pointed out that Vacca's formula may be deduced from the formula

$$\gamma = 1 - \int_0^1 \frac{F(x)}{1 + x} \, dx,$$

found by Catalan (1875) where $F(x) = \sum_{n=1}^{\infty} x^{2^n}$, since Vacca's formula can be written in the form

$$\gamma = \int_0^1 \left(\frac{x - x^3}{1 + x} + 2 \frac{x^3 - x^7}{1 + x} + 3 \frac{x^7 - x^{15}}{1 + x} + \cdots \right) dx$$

$$= \int_0^1 \frac{F(x)}{x(1 + x)} \, dx$$

and

$$\int_0^1 \frac{F(x)}{x} \, dx = \sum_{n=1}^{\infty} \frac{1}{2^n} = 1.$$

The power series $F(x)$ satisfies a simple functional equation

$$F(x) = x^2 + F(x^2),$$

which is a typical example of Mahler's functional equations in the theory of transcendental numbers.

Infinite Series 29

SOLUTION 2.11

Since it is easily verified that $\sin(2n+1)\theta$ is a polynomial of $\sin\theta$ by induction, we can write

$$p_n(\sin^2\theta) = \frac{\sin(2n+1)\theta}{\sin\theta}$$

where $p_n(x)$ is a polynomial of degree n satisfying $p_n(0) = 2n + 1$. The zeros of $p_n(x)$ can be obtained by solving the equation $\sin(2n+1)\theta = 0$ with $\sin\theta \neq 0$; therefore we get the following n points:

$$\sin^2\xi_{1,n} < \sin^2\xi_{2,n} < \cdots < \sin^2\xi_{n,n}$$

in the interval $(0, 1)$ where $\xi_{k,n} = k\pi/(2n + 1) \in (0, \pi/2)$. Hence we have

$$\sin(2n+1)\theta = (2n+1)\sin\theta \prod_{k=1}^{n}\left(1 - \frac{\sin^2\theta}{\sin^2\xi_{k,n}}\right).$$

By the substitution $x = (2n + 1)\theta/\pi$,

$$\frac{\sin\pi x}{\pi x} \cdot \frac{x_n}{\sin x_n} = \prod_{k=1}^{n}\left(1 - \frac{\sin^2 x_n}{\sin^2\xi_{k,n}}\right)$$

where $x_n = \pi x/(2n + 1)$.

We can assume x is not an integer; otherwise the expansion clearly holds. Take any positive integers n and m satisfying $n > m > |x|$ so that $|x_n| < \xi_{m,n}$. Putting

$$\eta_{m,n} = \prod_{k=m+1}^{n}\left(1 - \frac{\sin^2 x_n}{\sin^2\xi_{k,n}}\right),$$

we get

$$\lim_{n\to\infty}\frac{1}{\eta_{m,n}} = \frac{\pi x}{\sin\pi x}\prod_{k=1}^{m}\left(1 - \frac{x^2}{k^2}\right).$$

On the other hand,

$$1 > \eta_{m,n} \geq 1 - \sum_{k=m+1}^{n}\frac{\sin^2 x_n}{\sin^2\xi_{k,n}}$$

since

$$(1 - \alpha_1)\cdots(1 - \alpha_n) \geq 1 - \alpha_1 - \cdots - \alpha_n$$

Problems and Solutions in Real Analysis

for any $0 < \alpha_k < 1$ and any positive integer N. Using now the inequalities

$$\frac{2\theta}{\pi} < \sin \theta < \theta$$

holding for $0 < \theta < \pi/2$,

$$\eta_{m,n} \geq 1 - \frac{\pi^2}{4} \sum_{k=m+1}^{n} \frac{x_n^2}{\xi_{k,n}^2} \geq 1 - \frac{\pi^2 x^2}{4m},$$

since

$$\sum_{k=m+1}^{\infty} \frac{1}{k^2} < \frac{1}{m}.$$

Hence $\lim\limits_{n\to\infty} \eta_{m,n}$ converges to 1 as $m \to \infty$. $\qquad\square$

Chapter 3

Continuous Functions

- A function $f(x)$ defined on an interval $I \subset \mathbb{R}$ is said to be *uniformly continuous* on I provided that for any $\epsilon > 0$ there exists a positive number δ such that $|f(x) - f(x')| < \epsilon$ whenever $x, x' \in I$ and $|x - x'| < \delta$.

- A sequence of functions $\{f_n(x)\}$ defined on an open interval I is said to converge to $f(x)$ *uniformly on compact sets* provided that it converges uniformly on any compact set K contained in I; that is,

$$\sup_{x \in K} |f_n(x) - f(x)|$$

converges to 0 as $n \to \infty$.

- If a sequence of continuous functions converges to $f(x)$ uniformly on compact sets in I, then $f(x)$ is continuous on I.

The pointwise convergence of a sequence of continuous functions does not imply the continuity of the limit function in general. For example, Dirichlet's function

$$\lim_{n \to \infty} \left(\lim_{m \to \infty} (\cos(n!\,\pi x))^m \right)$$

takes the value 1 at every rational point x and 0 at every irrational point.

- Let $\{f_n(x)\}$ be a sequence of continuous functions defined on an interval I, converging pointwise to the limit function $f(x)$. R. Baire (1874–1932) showed that the set of continuity points of $f(x)$ is dense in I.

- The image of an interval under a continuous function is also an interval.

- A point x_0 is said to be a *discontinuity point of the first kind* of $f(x)$ provided that both right and left limits exist but are different from each other.

32   *Problems and Solutions in Real Analysis*

- A function $f(x)$ is said to be *piecewise continuous* if it is continuous except for a finite set of discontinuity points of the first kind.

- $f(x)$ defined on an interval I is said to be a *Lipschitz function* with constant $L > 0$ provided that

$$|f(x) - f(y)| \le L|x - y|$$

for all x and y in I. Clearly any Lipschitz function is uniformly continuous on I. The least constant L with which f satisfies the above Lipschitz condition is said to be the Lipschitz constant of f.

PROBLEM 3.1

Suppose that $f \in C(\mathbb{R})$ and that $f(x + 1) - f(x)$ converges to 0 as $x \to \infty$. Then show that

$$\frac{f(x)}{x}$$

also converges to 0 as $x \to \infty$. Suppose further that $f(x + y) - f(x)$ converges to 0 as $x \to \infty$ for an arbitrary fixed y. Show then that this convergence is uniform on compact sets in $\mathbb{R}$.

PROBLEM 3.2

Let $p_n(x)$ be any polynomial with integer coefficients whose degree is greater than or equal to 1, and let I be any closed interval of length ≥ 4. Then show that there exists at least one point x in I satisfying

$$|p_n(x)| \ge 2.$$

PROBLEM 3.3

Let $f_n \in C[a, b]$ be a monotone increasing sequence

$$f_1(x) \le f_2(x) \le \cdots,$$

which converges pointwise to $f(x) \in C[a, b]$. Show that the convergence is uniform on $[a, b]$.

PROBLEM 3.4

Suppose that $f \in C[0, \infty)$ and that $f(nx)$ converges to 0 as $n \to \infty$ for an arbitrary non-negative x. Prove or disprove that $f(x)$ converges to 0 as $x \to \infty$.

PROBLEM 3.5

Show that there are no continuous functions f, g and h defined on $\mathbb{R}$ satisfying

$$h(f(x) + g(y)) = xy$$

for all points (x, y) in $\mathbb{R}^2$.

PROBLEM 3.6

First put $E_1 = \{0, 1\}$ and suppose that a finite sequence $E_n \subset [0, 1]$ is given. Define E_{n+1} by inserting new fractions $(a + c)/(b + d)$ between every two consecutive fractions a/b and c/d in E_n. Of course we understand $0 = 0/1$ and $1 = 1/1$; thus,

$$E_2 = \left\{0, \frac{1}{2}, 1\right\}, \quad E_3 = \left\{0, \frac{1}{3}, \frac{1}{2}, \frac{2}{3}, 1\right\}, \quad \dots$$

With the help of the sequence E_n we define the piecewise linear continuous function $\phi_n(x)$ on the interval $[0, 1]$ such that

$$\phi_n\left(\frac{a}{b}\right) = \phi_n\left(\frac{c}{d}\right) = 0, \quad \phi_n\left(\frac{a + c}{b + d}\right) = \frac{1}{b + d}$$

and $\phi_n(x)$ is linear on the intervals

$$\left[\frac{a}{b}, \frac{a + c}{b + d}\right] \quad and \quad \left[\frac{a + c}{b + d}, \frac{c}{d}\right]$$

respectively, for every two successive terms a/b and c/d in E_n.

Show then that the series

$$f(x) = \sum_{n=1}^{\infty} \phi_n(x)$$

converges at every point $x \in [0, 1]$; more precisely, it converges to $1 - 1/q$ at any rational point $x = p/q \in [0, 1]$ with $(p, q) = 1$ and to 1 at any irrational point x in $(0, 1)$.

This problem arises in the author's paper (1995). The sequence of all reduced fractions with denominators not exceeding n, listed in order of their size, is called the *Farey sequence* of order n. See, for example, Niven and Zuckerman (1960) for details. It is easily seen that $f(x)$ is continuous at x if and only if x is irrational. This implies that the series $\sum \phi_n(x)$ never converge uniformly on any subinterval of $[0, 1]$.

PROBLEM 3.7

Let $c_1, c_2, ..., c_n$ and $\lambda_1, \lambda_2, ..., \lambda_n$ be real numbers with $\lambda_j \neq \lambda_k$ for any $j \neq k$. Show that $c_1 = c_2 = \cdots = c_n = 0$ if

$$\sum_{k=1}^{n} c_k \exp(\lambda_k i x)$$

converges to 0 as $x \to \infty$.

Solutions for Chapter 3

SOLUTION 3.1

For any $\epsilon > 0$ there exists an integer N satisfying

$$-\epsilon < f(x + 1) - f(x) < \epsilon$$

for any $x > N$. Summing the following $\ell = [x] - N$ inequalities

$$-\epsilon < f(x - j + 1) - f(x - j) < \epsilon$$

for $j = 1, ..., \ell$ and for $x \geq N + 1$ we get

$$-\epsilon([x] - N) < f(x) - f(x - \ell) < \epsilon([x] - N).$$

Since $N \leq x - \ell < N + 1$ it follows from this that

$$-M - \epsilon(x - N + 1) < f(x) < M + \epsilon(x - N),$$

where M is the maximum of $|f(x)|$ on $[N, N + 1]$. Therefore

$$\left| \frac{f(x)}{x} \right| < \frac{M + \epsilon(x - N + 1)}{x} < \epsilon + \frac{M}{x},$$

which implies that $|f(x)/x| < 2\epsilon$ for any $x > \max\{N + 1, M/\epsilon\}$.

To show the latter half of the problem put

$$g_x(y) = \sup_{t \geq x} |f(t + y) - f(t)|,$$

which converges to 0 as $x \to \infty$ for an arbitrary fixed y. If $g_x(y_0) > s$, then there exists $t_0 \geq x$ satisfying $|f(t_0 + y_0) - f(t_0)| > s$. By the continuity of f we have $|f(t + y) - f(t)| > s$ for any (t, y) sufficiently close to (t_0, y_0). This means that $g_x(y) > s$ for any y sufficiently close to y_0 ; in other words, the set

$$\{y \in \mathbb{R} ; g_x(y) > s\}$$

is an open set. Hence $\{g_n(y)\}$ is a sequence of Borel measurable functions converging pointwise to 0. By Egoroff's theorem we can find a measurable set F in $[-1, 1]$ whose measure is greater than $3/2$ such that $g_n(y)$ converges to 0 uniformly on F; therefore for any $\epsilon > 0$ there exists an integer N satisfying $g_n(y) < \epsilon$ for any $n > N$ and any y in F. By the theorem due to Steinhaus (1920) there exists an interval I such that any point y in I can be expressed as $y = u - v$ with $u, v \in F \cap (-F)$, since the measure of $F \cap (-F)$ is positive. Clearly $-y$ has such an expression, and we can assume that I is contained in the positive real axis. For any non-negative x we have

$$g_n(x + x') = \sup_{t \geq n} |f(t + x + x') - f(t)|$$

$$\leq \sup_{t \geq n} |f(t + x) - f(t)| + \sup_{t \geq n} |f(t + x + x') - f(t + x)|$$

$$= g_n(x) + g_{n+x}(x') \leq g_n(x) + g_n(x').$$

Applying the above inequality to $y = u - v \in I$ we get

$$g_n(y) = g_n(u - v) \leq g_n(u) + g_n(-v) < 2\epsilon,$$

since the case does not occur in which both u and $-v$ are negative. This implies that $\{g_n(y)\}$ converges to 0 uniformly on the interval I. This is also true on any set consisting of finite parallel translations of I since $g_n(x + x') \leq g_n(x) + g_n(x')$ for any $x \in I$ and x'.   □

SOLUTION 3. 2

Let I be any closed interval of length 4. We solve this problem for any polynomial

$$p_n(x) = a_n x^n + a_{n-1} x^{n-1} + \cdots + a_0$$

with a non-zero integer a_n and real $a_{n-1}, ..., a_0$. Since a_n is invariant under parallel translation, we can assume that $I = [-2, 2]$. Let M be the difference of the maximum and the minimum of p_n on the interval I. It may be convenient to introduce the notation

$$\sum_{k=0}^{n}{}^{*} b_k = b_0 + 2b_1 + 2b_2 + \cdots + 2b_{n-1} + b_n.$$

Now for any integer $0 \leq s < n$ put $\omega = -e^{s\pi i/n} \neq 1$; hence

$$\sum_{k=0}^{n}{}^{*} \omega^k = \frac{(1 + \omega)(1 - \omega^n)}{1 - \omega} = (1 - (-1)^{s+n}) \frac{1 + \omega}{1 - \omega}.$$

Since $\overline{\omega} = \omega^{-1}$, it can be seen that the real part of the above expression vanishes; in other words,

$$\sum_{k=0}^{n}{}^{*} (-1)^k \cos \frac{ks}{n} \pi = 0.$$

On the other hand, it is clear that $2\cos s\theta = e^{si\theta} + e^{-si\theta}$ is a polynomial in $2\cos\theta = e^{i\theta} + e^{-i\theta}$ with integer coefficients of degree s. Hence we can write

$$2\cos s\theta = \tau_s(2\cos\theta).$$

Since s is arbitrary, we get

$$\sum_{k=0}^{n}{}^{*}(-1)^k \alpha_k^s = 0$$

for any $0 \le s < n$, where

$$\alpha_k = 2\cos\frac{k\pi}{n}.$$

Also for $s = n$,

$$\sum_{k=0}^{n}{}^{*}(-1)^k \alpha_k^n = \sum_{k=0}^{n}{}^{*}(-1)^k \tau_n(\alpha_k)$$

$$= 2\sum_{k=0}^{n}{}^{*}(-1)^k \cos k\pi = 4n,$$

since the coefficient of the leading term of $\tau_n(x)$ is equal to 1. Hence we have

$$4n|a_n| = \left|\sum_{k=0}^{n}{}^{*}(-1)^k p_n(\alpha_k)\right|$$

$$\le \sum_{k=0}^{n-1} |p_n(\alpha_k) - p_n(\alpha_{k+1})| \le nM,$$

which implies that $M \ge 4|a_n| \ge 4$. We thus have $\max_{x\in I}|p_n(x)| \ge 2$.   □

REMARK. The maximum of $|\tau_n(x)|$ on the interval $[-2, 2]$ is clearly equal to 2, which means that we cannot replace 2 by any larger constant in general.

$$T_n(x) = \frac{1}{2}\tau_n(2x)$$

is a polynomial with integer coefficients of degree n and these form a system of orthogonal polynomials over the interval $[-1, 1]$. $T_n(x)$ is called the nth Chebyshev polynomial of the first kind and satisfies the relation

$$T_n(\cos\theta) = \cos n\theta.$$

See Chapter 15 for various properties on the Chebyshev polynomials.

$\boxed{\textbf{Solution 3.3}}$

For any $\epsilon > 0$ define the set

$$E_n = \{x \in [a, b] ; f(x) - f_n(x) \geq \epsilon\}.$$

Then $\{E_n\}$ is a sequence of monotone decreasing compact sets in view of the continuity of f_n and f. Suppose now that E_n is not empty for any positive integer n. It then follows that

$$\bigcap_{n=1}^{\infty} E_n \neq \emptyset.$$

Let x_0 be some point belonging to all the sets E_n. But this means that $\{f_n(x_0)\}$ does not converge to $f(x_0)$, contrary to the assumption. Thus E_n is empty for all sufficiently large n; in other words, $|f(x) - f_n(x)| < \epsilon$ for any $x \in [a, b]$. $\square$

$\boxed{\textbf{Solution 3.4}}$

We prove that the assertion is true. Suppose, on the contrary, that $f(x)$ does not converge to 0 as $x \to \infty$. We then find a strictly monotone increasing sequence $1 < x_1 < x_2 < \cdots$ diverging to ∞ and a positive constant δ satisfying

$$|f(x_k)| > 2\delta$$

for any positive integer k. By the continuity of f we can find a sufficiently small $\epsilon_k > 0$ such that $|f(x)| \geq \delta$ holds on the interval $[x_k - \epsilon_k, x_k + \epsilon_k]$ for each k. Now put

$$E_n = \bigcup_{k=n}^{\infty} \bigcup_{m=-\infty}^{\infty} \left(\frac{m - \epsilon_k}{x_k}, \frac{m + \epsilon_k}{x_k} \right)$$

for all positive integer n. E_n is an open and dense set since x_k diverges to ∞ as $k \to \infty$. Since $\mathbb{R}$ is a Baire space, the intersection

$$\bigcap_{n=1}^{\infty} E_n$$

is also a dense set; thus we can choose a point $x^* > 1$ which belongs to all the sets E_n. Namely there exist two integers $k_n \geq n$ and m_n satisfying

$$\left| x^* - \frac{m_n}{x_{k_n}} \right| < \frac{\epsilon_{k_n}}{x_{k_n}}$$

for all n. Note that m_n diverges to ∞ as $n \to \infty$. Therefore

$$\left| x_{k_n} - \frac{m_n}{x^*} \right| < \frac{\epsilon_{k_n}}{x^*} < \epsilon_{k_n},$$

which implies that $|f(m_n/x^*)| \geq \delta$, contrary to the assumption that $f(nx)$ converges to 0 as $n \to \infty$ at $x = 1/x^*$.   □

SOLUTION 3.5

If continuous functions f, g and h on $\mathbb{R}$ satisfy the relation

$$h(f(x) + g(y)) = xy$$

for all x and y, the function $h(x)$ must be surjective onto $\mathbb{R}$ since xy takes all real values. Now if $f(x) = f(x')$, then

$$x = h(f(x) + g(1)) = h(f(x') + g(1)) = x'.$$

Thus $f(x)$ is one-to-one and hence strictly monotone. Suppose that f is bounded above. The limit of $f(x)$ as $x \to \infty$ exists, say α. We then see that

$$h(\alpha + g(1)) = \lim_{x \to \infty} h(f(x) + g(1)) = \lim_{x \to \infty} x = \infty,$$

a contradiction. Thus $f(x)$ is unbounded above. The similar argument can be applied when $x \to -\infty$. Therefore f is one-to-one onto $\mathbb{R}$. Since

$$h(f(x) + g(0)) = 0$$

holds for all x, the function $h(x)$ vanishes identically. This is clearly a contradiction.   □

SOLUTION 3.6

Let p/q be any fraction in the interval $[0, 1]$. The fraction p/q is contained in E_q; otherwise, there exist two adjacent fractions a/b and c/d in E_q satisfying

$$\frac{a}{b} < \frac{p}{q} < \frac{c}{d},$$

which implies

$$\frac{1}{bd} = \frac{c}{d} - \frac{a}{b} = \frac{c}{d} - \frac{p}{q} + \frac{p}{q} - \frac{a}{b} \geq \frac{1}{dq} + \frac{1}{bq} \geq \frac{b+d}{bdq} > \frac{1}{bd},$$

a contradiction. We used here the facts that $bc - ad = 1$ and $b + d \geq n + 1$ for any two consecutive fractions a/b and c/d in E_n, which can be easily verified by induction on n.

Problems and Solutions in Real Analysis

We next show that

$$f\left(\frac{a}{b}\right) = 1 - \frac{1}{b} \tag{3.1}$$

for any fraction a/b belonging to E_n by induction on n. Clearly it holds for $a/b = 0/1$ and $1/1$ since $\phi_m(0) = \phi_m(1) = 0$ for all positive integer m. Suppose (3.1) holds for any a/b in E_n. By definition we have

$$f\left(\frac{a}{b}\right) = \phi_1\left(\frac{a}{b}\right) + \cdots + \phi_{b-1}\left(\frac{a}{b}\right) = 1 - \frac{1}{b}$$

since $a/b \in E_k$ for all $k \geq b$. Let a/b and c/d be any successive fractions in E_n and consider the fraction $(a+c)/(b+d) \in E_{n+1}$. Since

$$\Phi(x) = \phi_1(x) + \cdots + \phi_{n-1}(x)$$

is a linear function on the interval $[a/b, c/d]$, we get

$$\Phi\left(\frac{a+c}{b+d}\right) = 1 - \frac{1}{b} + \frac{1/b - 1/d}{c/d - a/b}\left(\frac{a+c}{b+d} - \frac{a}{b}\right)$$

$$= 1 - \frac{2}{b+d}.$$

Hence

$$f\left(\frac{a+c}{b+d}\right) = \Phi\left(\frac{a+c}{b+d}\right) + \phi_n\left(\frac{a+c}{b+d}\right) = 1 - \frac{1}{b+d}$$

by definition. Therefore (3.1) holds for any rational point in the interval $[0, 1]$.

Let m be any positive integer. Since the piecewise linear function $\Phi_m(x)$ takes the value less than 1 at any point belonging to E_{m+1}, as is already seen above, it follows clearly that $\Phi_m(x) < 1$ for all $x \in [0, 1]$. Thus the series

$$\sum_{n=1}^{\infty} \phi_n(x)$$

converges at every irrational point x. Let a/b and c/d be any adjacent fractions in E_{m+1} satisfying $a/b < x < c/d$. We then have

$$\Phi_m(x) \geq \min\left\{\Phi_m\left(\frac{a}{b}\right), \Phi_m\left(\frac{c}{d}\right)\right\}$$

$$= \min\left\{1 - \frac{1}{b}, 1 - \frac{1}{d}\right\}.$$

Since x is irrational, both b and d must diverge to ∞ as $m \to \infty$; so, $f(x) = 1$, as required. □

$\boxed{\text{SOLUTION 3.7}}$

Put

$$f(x) = \sum_{k=1}^{n} c_k \exp(\lambda_k i x).$$

For any $\epsilon > 0$ we can find a sufficiently large integer N satisfying $|f(x)| < \epsilon$ for all x greater than N. For each $1 \le k \le n$ we have

$$\frac{1}{T} \int_{T}^{2T} f(x) \exp(-\lambda_k i x)\, dx$$

$$= c_k + \frac{1}{T} \sum_{\ell \ne k} c_\ell \int_{T}^{2T} \exp((\lambda_\ell - \lambda_k) i x)\, dx$$

$$= c_k + \frac{1}{T} \sum_{\ell \ne k} c_\ell \frac{\exp(2(\lambda_\ell - \lambda_k) i T) - \exp((\lambda_\ell - \lambda_k) i T)}{(\lambda_\ell - \lambda_k) i}.$$

Therefore

$$|c_k| \le \frac{1}{T} \int_{T}^{2T} |f(x)|\, dx + \frac{2}{T} \sum_{\ell \ne k} \frac{|c_\ell|}{|\lambda_\ell - \lambda_k|}$$

$$< \epsilon + O\left(\frac{1}{T}\right)$$

for any $T > N$, which implies $c_k = 0$ since ϵ is arbitrary. $\square$

Chapter 4

Differentiation

- If $f(x)$ is continuous on a closed interval $[a, b]$ and differentiable on the open interval (a, b), then there exists a point c in (a, b) satisfying

$$f'(c) = 0$$

whenever $f(a) = f(b)$. This is known as *Rolle's theorem*, which is equivalent to the following *mean value theorem*: there exists a point c in (a, b) satisfying

$$\frac{f(b) - f(a)}{b - a} = f'(c)$$

whatever $f(a)$ and $f(b)$.

- If $f(x)$ and $g(x)$ are continuous on the closed interval $[a, b]$ and differentiable on the open interval (a, b) with $g(a) \neq g(b)$, and if $f'(x)$ and $g'(x)$ never vanish for the same value of x, then there exists a point $c \in (a, b)$ satisfying

$$\frac{f(b) - f(a)}{g(b) - g(a)} = \frac{f'(c)}{g'(c)}.$$

This is known as *Cauchy's mean value theorem*.

- For any differentiable function $f(x)$ on an interval I, the image of the derivative $f'(I)$ is always an interval. In other words, if $f(x)$ is differentiable on $[a, b]$, $f'(a) = \alpha$, $f'(b) = \beta$, and if η lies between α and β, then there is a ξ in (a, b) for which $f'(\xi) = \eta$. This is the theorem due to J. G. Darboux (1842–1917).

- If $f(x)$ is n times differentiable on an open interval I, then for a fixed a in I,

$$f(x) = f(a) + \frac{f'(a)}{1!}(x - a) + \cdots + \frac{f^{(n-1)}(a)}{(n-1)!}(x - a)^{n-1} + R_n.$$

This is called *Taylor's formula* and R_n is called the remainder term.

- J. L. Lagrange (1736–1813) showed that

$$R_n = \frac{f^{(n)}(c)}{n!}(x-a)^n$$

  for some c in I, known as Lagrange's form of the remainder term.

- Moreover if $f^{(n)}(x)$ is continuous on I, then the remainder term can be expressed in the integral form

$$R_n = \frac{1}{(n-1)!} \int_a^x (x-t)^{n-1} f^{(n)}(t)\, dt.$$

PROBLEM 4.1

Suppose that all roots of the algebraic equation $x^n + a_{n-1}x^{n-1} + \cdots + a_0 = 0$ have negative real parts and that $f \in C^n(0, \infty)$. Show that if

$$f^{(n)}(x) + a_{n-1}f^{(n-1)}(x) + \cdots + a_0 f(x)$$

converges to 0 as $x \to \infty$, then $f^{(k)}(x)$ also converges to 0 as $x \to \infty$ for any $0 \le k \le n$.

This does not hold if the algebraic equation has a root ξ with non-negative real part, because $e^{\xi x}$ is a solution of the differential equation

$$f^{(n)}(x) + a_{n-1}f^{(n-1)}(x) + \cdots + a_0 f(x) = 0$$

and does not converge to 0.

PROBLEM 4.2

Show that any $f \in C^2(\mathbb{R})$ satisfies the inequality

$$\left(\sup_{x\in\mathbb{R}} |f'(x)| \right)^2 \le 2 \sup_{x\in\mathbb{R}} |f(x)| \cdot \sup_{x\in\mathbb{R}} |f''(x)|.$$

Prove moreover that the constant 2 on the right-hand side cannot in general be replaced by any smaller number.

This was proved by Hadamard (1914) and was generalized by Kolmogorov (1939) to the inequality for $f \in C^n(\mathbb{R})$:

$$\left(\sup_{x \in \mathbb{R}} \left| f^{(k)}(x) \right| \right)^n \leq C_{k,n} \left(\sup_{x \in \mathbb{R}} |f(x)| \right)^{n-k} \left(\sup_{x \in \mathbb{R}} \left| f^{(n)}(x) \right| \right)^k$$

for $0 < k < n$ with the best possible constant $C_{k,n}$, which is a rational number expressible in terms of the Euler numbers. de Boor and Schoenberg (1976) gave a proof using spline functions. We present first several values of $C_{k,n}$:

$n \backslash k$	1	2	3	4
2	2			
3	$\dfrac{9}{8}$	3		
4	$\dfrac{512}{375}$	$\dfrac{36}{25}$	$\dfrac{24}{5}$	
5	$\dfrac{1953125}{1572864}$	$\dfrac{125}{72}$	$\dfrac{225}{128}$	$\dfrac{15}{2}$

Landau (1913) showed that any $f \in C^2(0, \infty)$ satisfies

$$\left(\sup_{x>0} |f'(x)| \right)^2 \leq 4 \sup_{x>0} |f(x)| \cdot \sup_{x>0} |f''(x)|$$

with the best possible constant 4. An explicit formula for general $C_{k,n}$ in this case is not known.

PROBLEM 4.3

Let $Q_n(x)$ be a polynomial with real coefficients of degree n and M be the maximum of $|Q_n(x)|$ on the interval $[-1, 1]$. Show that

$$\sqrt{1 - x^2} \, |Q'_n(x)| \leq nM$$

for any $-1 \leq x \leq 1$. Show next that

$$|Q'_n(x)| \leq n^2 M$$

for any $-1 \leq x \leq 1$.

The latter is called Markov's inequality, which first appeared in A. A. Markov (1889). Markov is famous for his study of Markov chains. The equality occurs for Chebyshev polynomial $T_n(x)$ of the first kind.

Concerning higher derivatives V. A. Markov (1892), a younger brother of A. A. Markov, showed that

$$\max_{-1 \le x \le 1} \left| Q_n^{(k)}(x) \right| \le \frac{n^2(n^2 - 1^2) \cdots (n^2 - (k-1)^2)}{1 \cdot 3 \cdot 5 \cdots (2k-1)} \max_{-1 \le x \le 1} |Q_n(x)|.$$

Note that the coefficient of $\| Q_n \|$ on the right-hand side is equal to $T_n^{(k)}(1)$ (See **PROBLEM 15.7**). V. A. Markov's paper was published when he was 21 years old, a student of St. Petersburg Univ., who died at the age of 25. Duffin and Schaeffer (1941) gave another proof for this. Rogosinski (1955) discussed this problem using only the classical Lagrange interpolation polynomials.

PROBLEM 4.4

Suppose that $f \in C^\infty(\mathbb{R})$ satisfies $f(0)f'(0) \ge 0$ and that $f(x)$ converges to 0 as $x \to \infty$. Show then that there exists an increasing sequence $0 \le x_1 < x_2 < x_3 < \cdots$ satisfying

$$f^{(n)}(x_n) = 0.$$

PROBLEM 4.5

Show that any $f \in C^{n+1}[0, 1]$ satisfies

$$\max_{0 \le x \le 1} \left| f^{(n+1)}(x) \right| \ge 4^n n!$$

if $f(0) = f'(0) = \cdots = f^{(n)}(0) = f'(1) = \cdots = f^{(n)}(1) = 0$ and $f(1) = 1$.

PROBLEM 4.6

Show that the maximum of $|Q^{(n)}(x)|$ over $[-1, 1]$ is equal to $2^n n!$ where

$$Q(x) = (1 - x^2)^n.$$

Multiplying $Q^{(n)}(x)$ by

$$\frac{1}{(-2)^n n!}$$

one gets the nth Legendre polynomial $P_n(x)$. See Chapter 14 for various properties of the Legendre polynomials.

Differentiation   47

PROBLEM 4.7

Define the piecewise linear function

$$g(x) = \begin{cases} x & for \quad 0 \le x < 1/2, \\ 1 - x & for \quad 1/2 \le x < 1, \end{cases}$$

and extend it to $\mathbb{R}$ periodically. Show that

$$T(x) = \sum_{n=0}^{\infty} \frac{1}{2^n} g(2^n x)$$

is continuous but nowhere differentiable.

This function found by Takagi (1903) is a simpler example of a nowhere differentiable function than Weierstrass'

$$W(x) = \sum_{n=0}^{\infty} a^n \cos b^n \pi x.$$

Many types of nowhere differentiable continuous functions were reported after Weierstrass' discovery in 1874. Lerch (1888) examined various trigonometric series like $W(x)$. The Takagi function did not seem, however, well-known in European mathematical circles of those days. Takagi constructed his function using a dyadic expansion of x in the interval $[0, 1)$, which Cesàro (1906) also used to define many such functions. The piecewise linear function $g(x) = \text{dist}(x, \mathbb{Z})$ in the above problem was also used by Faber (1907) to define

$$f(x) = \sum_{n=1}^{\infty} \frac{1}{10^n} g\left(2^{n!} x\right).$$

Landsberg (1908) also discussed the problem using the function $g(x)$ and a dyadic expansion of x. Later van der Waerden (1930) found the same kind of function

$$f(x) = \sum_{n=1}^{\infty} \frac{1}{10^n} g(10^n x)$$

and eventually the Takagi function itself was refound by de Rham (1957).

PROBLEM 4.8

Suppose that $f \in C^1(0, \infty)$ is positive. Then show that for an arbitrary constant $a > 1$,

$$\liminf_{x \to \infty} \frac{f'(x)}{(f(x))^a} \le 0.$$

The example $f(x) = \exp(x^2)$ shows that the above inequality does not hold for $a = 1$.

PROBLEM 4.9

Suppose that $f \in C^2(0, \infty)$ converges to α as $x \to \infty$ and that

$$f''(x) + \lambda f'(x)$$

is bounded above for some constant λ. Then show that $f'(x)$ converges to 0 as $x \to \infty$.

Solutions for Chapter 4

$\boxed{\text{SOLUTION 4.1}}$

We first consider the case $n = 1$ and assume that $f(x) \in C^1(0, \infty)$ is a complex-valued function. Suppose that $f'(x) + zf(x)$ converges to 0 as $x \to \infty$ where z is a complex number with $\lambda = \Re z > 0$. Differentiating $g(x) = e^{zx} f(x)$ we have

$$g'(x) = e^{zx}(f'(x) + zf(x));$$

therefore $g'(x)e^{-zx}$ converges to 0 as $x \to \infty$. This means that for any $\epsilon > 0$ there exists an x_ϵ satisfying $|g'(x)|e^{-\lambda x} < \epsilon$ for any x greater than x_ϵ. Hence

$$e^{\lambda x}|f(x)| = |g(x)| \le |g(x_\epsilon)| + \int_{x_\epsilon}^{x} |g'(t)| \, dt$$

$$< e^{\lambda x_\epsilon}|f(x_\epsilon)| + \epsilon \int_{-\infty}^{x} e^{\lambda t} \, dt.$$

Since the last expression is equal to

$$e^{\lambda x_\epsilon}|f(x_\epsilon)| + \frac{\epsilon}{\lambda} e^{\lambda x},$$

we have

$$|f(x)| < e^{\lambda(x_\epsilon - x)}|f(x_\epsilon)| + \frac{\epsilon}{\lambda}.$$

Thus $|f(x)| < 2\epsilon/\lambda$ for all sufficiently large x.

We prove the $n + 1$ case by assuming the n case. Let $-\xi$ be a root of

$$x^{n+1} + a_n x^n + \cdots + a_0 = 0$$

with $\Re \xi > 0$. Since this polynomial is written as

$$(x + \xi)(x^n + b_{n-1}x^{n-1} + \cdots + b_0),$$

we get

$$f^{(n+1)}(x) + a_n f^{(n)}(x) + \cdots + a_0 f(x) = \phi'(x) + \xi \phi(x)$$

where

$$\phi(x) = f^{(n)}(x) + b_{n-1}f^{(n-1)}(x) + \cdots + b_0 f(x).$$

By the hypothesis $\phi(x)$ converges to 0 and hence so does $f^{(k)}(x)$ as $x \to \infty$ for each integer $0 \le k \le n$. Clearly $f^{(n+1)}(x)$ converges to 0 as well.   □

SOLUTION 4. 2

Put

$$\alpha = \sup_{x \in \mathbb{R}} |f(x)| \quad \text{and} \quad \beta = \sup_{x \in \mathbb{R}} |f''(x)|.$$

We may of course assume that both α and β are finite. If β vanishes, then α is finite if and only if $f(x)$ vanishes everywhere; therefore we can also assume that β is positive. For any x and positive y it follows from Taylor's formula that there is a $\xi_{x,y}$ satisfying

$$f(x + y) = f(x) + f'(x)y + f''(\xi_{x,y}) \frac{y^2}{2}.$$

Therefore we have

$$f(x + y) - f(x - y) = 2f'(x)y + \left(f''(\xi_{x,y}) - f''(\xi_{x,-y}) \right) \frac{y^2}{2},$$

which implies that

$$2|f'(x)|y = \left| f(x + y) - f(x - y) + \left(f''(\xi_{x,-y}) - f''(\xi_{x,y}) \right) \frac{y^2}{2} \right|$$
$$\le 2\alpha + \beta y^2.$$

Thus we get

$$\sup_{x \in \mathbb{R}} |f'(x)| \le \frac{\alpha}{y} + \frac{\beta y}{2},$$

where the right-hand side attains its minimum $\sqrt{2\alpha\beta}$ at $y = \sqrt{2\alpha/\beta}$.

To see that 2 is the best possible constant we first define an even step function

$$\phi''(x) = \begin{cases} 0 & \text{for} \quad |x| > 2, \\ 1 & \text{for} \quad 1 \le |x| \le 2, \\ -1 & \text{for} \quad |x| < 1. \end{cases}$$

Then

$$\phi'(x) = \int_{-2}^{x} \phi''(t)\,dt$$

is an odd piecewise linear continuous function and in turn

$$\phi(x) = \int_{-2}^{x} \phi'(t)\,dt - \frac{1}{2}$$

is an even C^1-function. The maxima of $|\phi(x)|$, $|\phi'(x)|$ and $|\phi''(x)|$ are clearly $1/2$, 1 and 1 respectively. The equality certainly holds in this example. However ϕ does not belong to $C^2(\mathbb{R})$. To tide over this difficulty it suffices to transform ϕ'' slightly to a continuous one in the neighborhood of the discontinuity points of ϕ'' so that its influence on ϕ and ϕ' becomes arbitrarily small. □

SOLUTION 4.3

The proof is based on Cheney (1966), p. 89−91. We first show that any polynomial $Q(x)$ with complex coefficients of degree $n - 1$ satisfies the inequality

$$\max_{-1 \le x \le 1} |Q(x)| \le n \max_{-1 \le x \le 1} \sqrt{1 - x^2}\,|Q(x)|.$$

Let M denote the right-hand side. If $x^2 \le 1 - 1/n^2$, then clearly $|Q(x)| \le M$. Hence we can assume that

$$|x| > \sqrt{1 - 1/n^2} > \cos \frac{\pi}{2n}.$$

The nth Chebyshev polynomial $T_n(x)$ of the first kind (See Chapter 15) is factorized as

$$T_n(x) = (x - \xi_1)\cdots(x - \xi_n)$$

where

$$\xi_k = \cos \frac{2k - 1}{2n}\pi$$

for $k = 1, 2, ..., n$. The Lagrange interpolation polynomial for Q with nodes $\xi_1, ..., \xi_n$ is

$$\sum_{k=1}^{n} \frac{Q(\xi_k)}{T_n'(\xi_k)} \cdot \frac{T_n(x)}{x - \xi_k} = \frac{1}{n} \sum_{k=1}^{n} (-1)^{k-1} Q(\xi_k) \sqrt{1 - \xi_k^2}\, \frac{T_n(x)}{x - \xi_k},$$

which is a polynomial of degree less than n; hence it coincides with the polynomial $Q(x)$. Using the fact that $\mathrm{sgn}(x - \xi_k)$ is independent of k in view of $\xi_1 < |x| \le 1$, we get

$$|Q(x)| \le \frac{M}{n^2} \sum_{k=1}^{n} \left| \frac{T_n(x)}{x - \xi_k} \right| = \frac{M}{n^2} \left| \sum_{k=1}^{n} \frac{T_n(x)}{x - \xi_k} \right| = \frac{M}{n^2} |T_n'(x)|.$$

Since

$$|T_n'(\cos\theta)| = n\left|\frac{\sin n\theta}{\sin\theta}\right| \le n^2,$$

we get $|Q(x)| \le M$, as required.

By the substitution $x = \cos\theta$ our inequality is equivalent to

$$\max_\theta |Q(\cos\theta)| \le n\max_\theta |\sin\theta\, Q(\cos\theta)|,$$

which is valid for all polynomial Q with complex coefficients of degree $n - 1$. Let $S(\theta)$ be a linear combination over $\mathbb{C}$ of 1, $\cos\theta$, $\cos 2\theta$, ..., $\cos n\theta$ and $\sin\theta$, $\sin 2\theta$, ..., $\sin n\theta$. For any ω and θ we put

$$S_0(\theta) = \frac{S(\omega + \theta) - S(\omega - \theta)}{2}.$$

Since $S_0(\theta)$ is an odd function, this is a linear combination of 1, $\sin\theta$, ..., $\sin n\theta$ only. Thus $S_0(\theta)/\sin\theta$ is a polynomial in $\cos\theta$ of degree less than n, since $\sin k\theta/\sin\theta$ can be expressed as a polynomial in $\cos\theta$ of degree $k - 1$. Applying our inequality to this polynomial in $\cos\theta$, we have

$$\max_\theta \left|\frac{S_0(\theta)}{\sin\theta}\right| \le n\max_\theta |S_0(\theta)| \le n\max_\theta |S(\theta)|.$$

Therefore

$$\lim_{\theta\to 0}\frac{S_0(\theta)}{\sin\theta} = \lim_{\theta\to 0}\frac{S_0(\theta)}{\theta} = S_0'(0) = S'(\omega),$$

which implies that

$$\max_\theta |S'(\theta)| \le n\max_\theta |S(\theta)|$$

since ω is arbitrary. This is called Bernstein's inequality (1912b).

Two inequalities stated in the problem can be solved by using Bernstein's inequality. Let $P(x)$ be any polynomial with complex coefficients of degree n. Then $P(\cos\theta)$ is a linear combination of 1, $\cos\theta$, ..., $\cos n\theta$ and it follows from Bernstein's inequality that

$$\max_{-1\le x\le 1}\sqrt{1 - x^2}\,|P'(x)| = \max_\theta |\sin\theta\, P'(\cos\theta)|$$

$$\le n\max_\theta |P(\cos\theta)| = nM.$$

Therefore, since $P'(x)$ is a polynomial of degree $n - 1$, we get

$$\max_{-1 \le x \le 1} |P'(x)| \le n \max_{-1 \le x \le 1} \sqrt{1 - x^2} \, |P'(x)|$$
$$\le n^2 M.$$

This completes the proof. $\qquad\qquad\qquad\qquad\qquad\qquad\qquad\qquad\qquad\qquad\qquad$ $\square$

$\boxed{\text{SOLUTION 4.4}}$

Suppose first that $f'(x)$ is positive for all $x \ge 0$. Then $f(x)$ is strictly monotone increasing and $f(0) \ge 0$ because $f'(0)$ is positive. This is a contradiction because $f(x)$ converges to 0 as $x \to \infty$. Similarly we get a contradiction, if $f'(x)$ is negative for all $x \ge 0$. Hence there exists at least one point $x_1 \ge 0$ at which $f'(x)$ vanishes.

Suppose now that we could find n points $x_1 < \cdots < x_n$ satisfying $f^{(k)}(x_k) = 0$ for $1 \le k \le n$. If $f^{(n+1)}(x)$ is positive for any $x > x_n$, then clearly $f^{(n)}(x) \ge f^{(n)}(x_n + 1) > 0$ for any $x \ge x_n + 1$, since $f^{(n)}(x_n) = 0$. Thus we have

$$f(x) \ge \frac{1}{n!} f^{(n)}(x_n + 1) \, x^n$$

$$+ \text{ (some polynomial of degree less than } n \text{)},$$

contrary to the assumption that $f(x)$ converges to 0. Similarly we would have a contradiction if $f^{(n+1)}(x)$ is negative for any $x > x_n$. Hence there exists at least one point x_{n+1} greater than x_n satisfying $f^{(n+1)}(x_{n+1}) = 0$. $\qquad$ $\square$

$\boxed{\text{SOLUTION 4.5}}$

Let

$$P(x) = x^n + a_{n-1} x^{n-1} + \cdots + a_0$$

be any polynomial with real coefficients. Integrating by parts repeatedly we have

$$\int_0^1 P(x) f^{(n+1)}(x) \, dx = -\int_0^1 P'(x) f^{(n)}(x) \, dx$$

$$\vdots$$

$$= (-1)^n \int_0^1 P^{(n)}(x) f'(x) \, dx = (-1)^n n!,$$

where we used $f(1) = 1$. Now taking P as the polynomial attaining the minimum in **PROBLEM 5.6** we get

$$
\begin{aligned}
n! = \left| \int_0^1 P(x) f^{(n+1)}(x)\,dx \right| &\le \max_{0 \le x \le 1} \left| f^{(n+1)}(x) \right| \int_0^1 |P(x)|\,dx \\
&= \frac{1}{4^n} \max_{0 \le x \le 1} \left| f^{(n+1)}(x) \right|.
\end{aligned}
$$

$\square$

SOLUTION 4.6

By Cauchy's integral formula we get

$$
Q^{(n)}(x) = \frac{n!}{2\pi i} \int_C \frac{(1 - z^2)^n}{(z - x)^{n+1}}\,dz
$$

where C is an oriented circle centered at $z = x$ with radius $r > 0$. Hence putting

$$
z = x + r e^{i\theta}
$$

for $0 \le \theta < 2\pi$ we obtain

$$
\begin{aligned}
Q^{(n)}(x) &= \frac{n!}{2\pi} \int_0^{2\pi} \frac{(1 - (x + r e^{i\theta})^2)^n}{r^{n+1} e^{(n+1)i\theta}}\, r e^{i\theta}\,d\theta \\
&= \frac{n!}{2\pi} \int_0^{2\pi} \left(\frac{1 - (x + r e^{i\theta})^2}{r e^{i\theta}} \right)^n d\theta.
\end{aligned}
$$

The expression in the parentheses can be written as

$$
\left(\frac{1 - x^2}{r} - r \right) \cos\theta - 2x - i\left(\frac{1 - x^2}{r} + r \right) \sin\theta.
$$

We now take $r = \sqrt{1 - x^2}$ for $|x| < 1$ so that $|P^{(n)}(x)| \le 2^n n!$. This inequality clearly holds for $x = 1$ and for $x = -1$. $\square$

SOLUTION 4.7

The continuity of $T(x)$ is obvious since it is defined as the series of continuous functions converging uniformly. To show the non-differentiability it suffices to consider any point x in the interval $(0, 1]$ since $T(x)$ is periodic with period 1.

We first consider any point x which can be expressed in the form $k/2^m$ with some odd integer k and non-negative integer m. For any integer $n \ge m$ put $h_n = 1/2^n$ for brevity. Then for any integer ℓ in $[0, n)$ there are no integers nor half-integers in the interval $(2^\ell x, 2^\ell(x + h_n))$. For if $2^\ell x < p/2 < 2^\ell x + 2^{\ell - n}$ for

some integer p, then we would have $2^n x = k2^{n-m} < 2^{n-\ell-1}p < k2^{n-m} + 1$, a
contradiction. This means that $g(x)$ is a linear function having the slope 1 or -1
on this subinterval. Hence

$$\frac{T(x + h_n) - T(x)}{h_n} = \sum_{\ell=0}^{\infty} \frac{g(2^\ell(x + h_n)) - g(2^\ell x)}{2^\ell h_n}$$

$$= \sum_{\ell=0}^{n-1} \frac{g(2^\ell x + 2^{\ell-n}) - g(2^\ell x)}{2^{\ell-n}}$$

is a finite sum of 1 or -1, which does not converge as $n \to \infty$.

Next consider any point x for which $2^n x$ is not an integer for all positive integer n. Since $2^n x$ is not an integer, we can find two positive numbers h_n and h'_n satisfying $[2^n x] = 2^n(x - h'_n)$ and $[2^n x] + 1 = 2^n(x + h_n)$. Note that $h_n + h'_n = 2^{-n}$. Then for any integer ℓ in $[0, n)$ there are no integers nor half-integers in the interval $(2^\ell(x - h'_n), 2^\ell(x + h_n))$. For if $p/2$ were contained in this interval for some integer p, then we have $[2^n x] < 2^{n-\ell-1}p < [2^n x] + 1$, a contradiction. Therefore

$$\frac{T(x + h_n) - T(x - h'_n)}{h_n + h'_n} = \sum_{\ell=0}^{\infty} \frac{g(2^\ell(x + h_n)) - g(2^\ell(x - h'_n))}{2^\ell(h_n + h'_n)}$$

$$= \sum_{\ell=0}^{n-1} \frac{g(2^\ell(x + h_n)) - g(2^\ell(x - h'_n))}{2^{\ell-n}}$$

is a finite sum of 1 or -1, which does not converge as $n \to \infty$.   □

SOLUTION 4.8

Suppose, contrary to the assertion, that there are positive numbers δ and x_0 satisfying

$$\delta < \frac{f'(x)}{(f(x))^a}$$

for any $x > x_0$. Then integrating from x_0 to x we have

$$\delta(x - x_0) < \int_{x_0}^{x} \frac{f'(t)}{(f(t))^a} dt$$

$$= \frac{1}{a - 1}\left(\frac{1}{(f(x_0))^{a-1}} - \frac{1}{(f(x))^{a-1}}\right).$$

Thus

$$\frac{1}{(f(x_0))^{a-1}} > \frac{1}{(f(x))^{a-1}} + (a-1)\delta(x-x_0)$$
$$> (a-1)\delta(x-x_0),$$

where the right-hand side diverges to ∞ together with x, giving a contradiction. $\square$

SOLUTION 4.9

The proof is substantially due to Hardy and Littlewood (1914). We use Taylor's formula with the integral remainder term

$$f(x+y) = f(x) + yf'(x) + y^2 \int_0^1 (1-t)f''(x+yt)\,dt,$$

valid for any $x > 1$ and $|y| < 1$. Now let us consider the integral on the right-hand side with f'' replaced by f'. By the mean value theorem there is a $\xi_{x,y}$ between 0 and y satisfying

$$y^2 \int_0^1 (1-t)f'(x+yt)\,dt = \int_x^{x+y} f(s)\,ds - yf(x)$$
$$= yf(x+\xi_{x,y}) - yf(x).$$

Since there exists a positive constant K satisfying $f''(x) + \lambda f'(x) \le K$, we have

$$f(x+y) - f(x) - yf'(x) + \lambda yf(x+\xi_{x,y}) - \lambda yf(x)$$
$$= y^2 \int_0^1 (1-t)\big(f''(x+yt) + \lambda f'(x+yt)\big)\,dt$$
$$\le Ky^2 \int_0^1 (1-t)\,dt = \frac{K}{2}y^2.$$

For the case in which $0 < y < 1$ we get

$$f'(x) \ge \frac{f(x+y) - f(x)}{y} + \lambda f(x+\xi_{x,y}) - \lambda f(x) - \frac{K}{2}y.$$

Making $x \to \infty$ we thus have

$$\liminf_{x\to\infty} f'(x) \ge -\frac{K}{2}y.$$

Therefore $\liminf_{x\to\infty} f'(x)$ must be ≥ 0 because y is arbitrary.

Differentiation 57

Similarly for the case in which $-1 < y < 0$,

$$f'(x) \leq \frac{f(x) - f(x - |y|)}{|y|} + \lambda f(x + \xi_{x,y}) - \lambda f(x) + \frac{K}{2}|y|,$$

which implies that

$$\limsup_{x \to \infty} f'(x) \leq \frac{K}{2}|y|$$

so that $\limsup_{x \to \infty} f'(x)$ is ≤ 0 because y is arbitrary. Therefore $f'(x)$ converges to 0 as $x \to \infty$. $\square$

Chapter 5

Integration

Let $f(x)$ be a bounded function defined on $I = [a, b]$. For given points $a = x_0 < x_1 < \cdots < x_{n-1} < x_n = b$ we divide I into n subintervals $I_k = [x_{k-1}, x_k]$. This division is denoted by Δ and the longest length of I_k is denoted by $|\Delta|$.

- The finite sum

$$\sum_{k=1}^{n} f(\xi_k)(x_k - x_{k-1}),$$

where each ξ_k is taken in I_k, is called the *Riemann sum* associated with the division Δ. In particular, if $x_k - x_{k-1} = 1/n$ for $1 \le k \le n$, the sum is called the equally divided Riemann sum.

- $f(x)$ is said to be *integrable in the sense of Riemann* on I provided that the Riemann sum associated with any division with arbitrarily chosen $\{\xi_k\}$ converges to a unique value as $|\Delta| \to 0$. This unique value is denoted by

$$\int_a^b f(x)\,dx.$$

- A bounded function $f(x)$ is integrable on I in the sense of Riemann if and only if the set of discontinuity points of f in I is a null set.

The term 'null set' is usually used in the theory of Lebesgue integration. However it is readily understandable independently; that is, a set X in $\mathbb{R}$ is said to be a null set provided that, for any $\epsilon > 0$, one can find a sequence of intervals $\{J_n\}$ satisfying

$$X \subset \bigcup_{n=1}^{\infty} J_n$$

and the total sum of the lengths of J_n is smaller than ϵ.

- A function $f(x)$ defined on $I = [a, b]$ is said to be of bounded variation on I if

$$\sup_{\Delta} |f(x_{i+1}) - f(x_i)|$$

  is finite, where the supremum is taken over all divisions of $[a, b]$. Any function of bounded variation is integrable in the sense of Riemann.

  If $f(x)$ is integrable over $[0, 1]$ in the sense of Riemann, then of course the equally divided Riemann sum converges:

$$\frac{1}{n} \sum_{k=1}^{n} f\left(\frac{k}{n}\right) \to \int_0^1 f(x)\,dx$$

  as $n \to \infty$. This is the usual definition of the definite integral for a continuous function. Then the following question naturally arises: Under what condition on a given sequence $\{a_n\} \subset [0, 1]$ can we assert that

$$\frac{1}{n} \sum_{k=1}^{n} f(a_k) \to \int_0^1 f(x)\,dx$$

  as $n \to \infty$ for every continuous function $f(x)$ on the interval $[0, 1]$? If you are interested in this question, you are certainly standing at the door of the theory of uniform distribution (See Chapter 12).

- Let $f(x)$ be continuous on $[a, b]$ and $g(x)$ be a non-negative integrable function on $[a, b]$ in the sense of Riemann. Then there exists $c \in (a, b)$ satisfying

$$\int_a^b f(x)g(x)\,dx = f(c) \int_a^b g(x)\,dx,$$

  which is called the first mean value theorem.

- Let $f(x)$ be a positive monotone decreasing function on $[a, b]$ and $g(x)$ be integrable in the sense of Riemann. Then there exists $c \in (a, b]$ satisfying

$$\int_a^b f(x)g(x)\,dx = f(a+) \int_a^c g(x)\,dx,$$

  which is called the second mean value theorem.

The remainder terms in rational approximations to some transcendental numbers can be expressed in a comparatively simple integral form. For example, one can find the following example in le Lionnais (1983):

$$\frac{22}{7} - \pi = \int_0^1 \frac{x^4(1-x)^4}{1+x^2}\,dx.$$

Zu Chongzhi (429–501) found two good rational approximations to π : 22/7 and 355/113. It therefore may be interesting to look for an analogous beautiful formula for $355/113 - \pi$. The latter is correct to 6 decimal places and Delahaye (1997) stated that this represents a precision that Europe had to wait until the 16th century.

Compare with **Problem 6.7** where a similar integral representation for the remainder term for Euler's constant is given, although we do not know whether it is transcendental or not.

Problem 5.1

Suppose that $f \in C[0, 1]$ and $g \in C(\mathbb{R})$ is a periodic function with period 1. Show then that

$$\lim_{n\to\infty} \int_0^1 f(x)g(nx)\,dx = \int_0^1 f(x)\,dx \int_0^1 g(x)\,dx.$$

As an example, take $g(x) = \sin 2\pi x$. Then

$$\lim_{n\to\infty} \int_0^1 f(x)\sin 2\pi nx\,dx = 0$$

for any continuous function $f(x)$ on the interval $[0, 1]$. This result valid for every integrable function $f(x)$ in the sense of Lebesgue, is known as the Riemann-Lebesgue lemma.

Problem 5.2

Find an example of a sequence of continuous functions $\{f_n\}$ defined on the interval $[0, 1]$ such that $0 \le f_n(x) \le 1$,

$$\int_0^1 f_n(x)\,dx$$

converges to 0 as $n \to \infty$, and that $\{f_n(x)\}$ does not converge at any point x in $[0, 1]$.

PROBLEM 5.3

Show that

$$\max_{0 \le x \le 1} |f'(x)| \ge 4 \int_0^1 |f(x)| \, dx$$

for any $f \in C^1[0, 1]$ satisfying $f(0) = f(1) = 0$. Prove moreover that 4 cannot be replaced by any larger constant.

PROBLEM 5.4

For any positive integer n show that

$$\int_0^1 |f(x)|^n |f'(x)| \, dx \le \frac{1}{n+1} \int_0^1 |f'(x)|^{n+1} \, dx$$

holds for any $f \in C^1[0, 1]$ satisfying $f(0) = 0$. Verify that the equality holds if and only if $f(x)$ is a linear function.

Opial (1960) showed this inequality for the case $n = 1$. The general case was proved by Hua (1965).

PROBLEM 5.5

Suppose that both $f(x)$ and $g(x)$ are monotone increasing continuous functions defined on $[0, 1]$. Show that

$$\int_0^1 f(x) \, dx \int_0^1 g(x) \, dx \le \int_0^1 f(x) g(x) \, dx.$$

According to Franklin (1885) Hermite stated this theorem in his course as communicated by Chebyshev .

PROBLEM 5.6

Show that the minimum of the integral

$$\int_0^1 |x^n + a_1 x^{n-1} + \cdots + a_n| \, dx$$

as $a_1, a_2, ..., a_n$ range over all real numbers, is equal to 4^{-n}.

PROBLEM 5.7

For any $f \in C^1[0, 1]$ show that

$$\sum_{k=1}^{n} f\left(\frac{k}{n}\right) - n \int_0^1 f(x)\, dx$$

converges to

$$\frac{f(1) - f(0)}{2}$$

as $n \to \infty$.

PROBLEM 5.8

Put $\phi(x) = 1/\sqrt{1 + |x|}$ and $\alpha_{j,n} = (j - 1/2)3^{-n}$ for $1 \le j \le 3^n$, $n \ge 0$.

(a) Show that there exist $\{c_{j,n}\}_{1 \le j \le 3^n, n \ge 0} \subset (0, 1)$ and $\{\lambda_n\}$ with $\lambda_n > 3^{5n+1}$ such that

$$\max_{0 \le x \le 1} \sum_{k=0}^{n} \psi_k(x) < 1 - \frac{1}{n + 2} \tag{5.1}$$

and

$$\sum_{k=0}^{n} \psi_k(\alpha_{j,n}) > 1 - \frac{1}{n + 1} \tag{5.2}$$

for any $1 \le j \le 3^n$ and $n \ge 0$, where

$$\psi_k(x) = \sum_{j=1}^{3^k} c_{j,k}\, \phi\left(\lambda_k(x - \alpha_{j,k})\right).$$

(b) Show that

$$\Psi(x) = \sum_{n=0}^{\infty} \int_0^x \psi_n(t)\, dt \tag{5.3}$$

is differentiable and satisfies $\Psi'(x) = \sum_{n=0}^{\infty} \psi_n(x)$ for any x in $(0, 1)$.

(c) Using the function Ψ construct an example of everywhere differentiable but nowhere monotone functions.

This is substantially due to Katznelson and Stromberg (1974). The first example of everywhere differentiable but nowhere monotone functions was constructed by Köpcke in a series of papers (1887, 89, 90) by graphical construction. Pereno (1897) gave a simpler example by a similar method, which is reproduced in the book of Hobson (1957), pp. 412–421.

This kind of examples illustrates a peculiar aspect of 'derivatives'. Darboux showed that every derivative must satisfy the intermediate value theorem, irrespective of whether it is continuous or not. Moreover Baire's theorem implies that the derivative has so many continuity points, since it is the limit of continuous functions

$$n\left(f\left(x+\frac{1}{n}\right)-f(x)\right)$$

as $n \to \infty$. So one may imagine that derivatives behave well. However the above example shows that the derivative may oscillate everywhere.

Some advanced readers may notice that the function $f(x)$ constructed in (c) is absolutely continuous. As a consequence, the derivative $f'(x)$ is not integrable in the sense of Riemann on any subinterval. To see this, suppose, on the contrary, that $f'(x)$ is continuous almost everywhere. Then $f'(x)$ vanishes almost everywhere, since it vanishes at every continuity point. On the other hand, since $f(x)$ is absolutely continuous, $f'(x)$ is integrable in the sense of Lebesgue; hence

$$f(x) = f(0) + \int_0^x f'(t)\,dt = f(0),$$

a contradiction.

PROBLEM 5.9

Prove that

$$\sum_{n=1}^{m} \frac{\sin n\theta}{n} < \int_0^\pi \frac{\sin x}{x}\,dx = 1.8519...$$

for any positive integer m and any θ in $[0,\pi]$. Show moreover that the constant on the right-hand side cannot be replaced by any smaller number.

This was conjectured by Fejér (1910) and proved by Jackson (1911) and by Gronwall (1912) independently. See also **PROBLEM 1.7** and **PROBLEM 7.10**.

In general, let $f(x)$ be a function of bounded variation defined on the interval $[-\pi,\pi]$ with $f(\pi) = f(-\pi)$ and $x = 0$ be a discontinuity point of f of the first kind satisfying $f(0+) = -f(0-) = 1$. It is then known that the nth partial sum $s_n(x)$ of the Fourier series

$$\frac{a_0}{2} + \sum_{n=1}^{\infty} (a_n \cos nx + b_n \sin nx)$$

converges to

$$\frac{f(x+) + f(x-)}{2}$$

as $n \to \infty$ for any x. Thus the convergence is not uniform [Why?]. Moreover it follows that

$$\lim_{n \to \infty} s_n\left(\frac{\pi}{n}\right) = \frac{2}{\pi} \int_0^{\pi} \frac{\sin x}{x} \, dx = 1.178979744472167....$$

This is known as the Gibbs phenomenon, in which the curves of $y = s_n(x)$ condense to a longer interval than $[-1, 1]$ on the y-axis. Gibbs (1899) reported it in 'Nature' in response to Michelson (1898), known for his work on the measurement of the speed of light in 1887 and as the first American to receive a Nobel prize in Physics in 1907.

Solutions for Chapter 5

SOLUTION 5.1

Without loss of generality we can replace $g(x)$ by $g(x) + c$ for any constant c; so we can assume that $g(x)$ is positive. By the periodicity of g

$$\int_0^1 f(x)g(nx)\,dx = \frac{1}{n}\int_0^n f\left(\frac{y}{n}\right)g(y)\,dy$$

$$= \frac{1}{n}\sum_{k=0}^{n-1}\int_k^{k+1} f\left(\frac{y}{n}\right)g(y)\,dy$$

$$= \frac{1}{n}\sum_{k=0}^{n-1}\int_0^1 f\left(\frac{k+s}{n}\right)g(s)\,ds.$$

Applying the first mean value theorem to each integral on the right-hand side, the above expression can be written as the product of $\int_0^1 g(s)\,ds$ times

$$\frac{1}{n}\sum_{k=0}^{n-1} f\left(\frac{k+s_k}{n}\right)$$

for some s_k in the interval $(0, 1)$, which is the Riemann sum converging to $\int_0^1 f(x)\,dx$ as $n \to \infty$.   □

REMARK. Note that this is also valid for any piecewise continuous function f. Moreover we can show that

$$\lim_{n\to\infty}\int_a^b f(x)g(nx)\,dx = \int_a^b f(x)\,dx\int_0^1 g(x)\,dx$$

for any continuous function f on a finite interval $[a, b]$. To see this take suitable integers $p < q$ satisfying $[a, b] \subset [p, q]$ and extend f on this wider interval $[p, q]$ by

$$\tilde{f}(x) = \begin{cases} 0 & \text{for}\quad p \le x < a, \\ f(x) & \text{for}\quad a \le x \le b, \\ 0 & \text{for}\quad b < x \le q. \end{cases}$$

Then apply the above result to $\tilde{f}$ and g on each interval $[k, k+1]$.

By the same argument we can generalize this formula to any $f \in C(a, b)$ when the improper integral

$$\int_a^b |f(x)| \, dx$$

converges. Indeed, apply the above formula to the interval $[a + \delta, b - \delta]$ for any $\delta > 0$ so that the improper integrals on the remainder intervals $(a, a + \delta]$ and $[b - \delta, b)$ are sufficiently small.

SOLUTION 5.2

Divide the unit interval $[0, 1]$ into $m \geq 3$ equal subintervals. For each subinterval I let I' and I'' be the left and right neighboring subinterval respectively with I' or I'' empty at endpoints. We associate with I the trapezoidal function $g_I(x)$ such that $g_I(x) = 1$ on I, $g_I(x) = 0$ on $[0, 1] \setminus (I \cup I' \cup I'')$ and that $g_I(x)$ is linear on I' and on I''. We thus have m continuous functions g_I's for each m. We call I the support of g_I. Arranging the g_I's in a line in any manner for $m = 3, 4, \ldots$ we define a sequence of continuous functions $\{f_n(x)\}$. It is clear that

$$a_n = \int_0^1 f_n(x) \, dx \leq \frac{2}{m}$$

if $f_n = g_I$ and the length of I is equal to $1/m$. Since $m \to \infty$ together with n, it follows that a_n converges to 0 as $n \to \infty$.

For any point x in $[0, 1]$ there are infinitely many cases in which x is contained in some support; in other words, $f_n(x) = 1$ for infinitely many n's. On the other hand, there are also infinitely many cases in which x is contained in neither the support nor its neighbors; in other words, $f_n(x) = 0$ for infinitely many n's. Hence $\{f_n(x)\}$ does not converge at any point x. $\qquad\square$

SOLUTION 5.3

Let $g(x)$ be any one of the four functions $f(x)$, $-f(x)$, $f(1-x)$ and $-f(1-x)$. Let α be the maximum of $|g'(x)|$ on the interval $[0, 1]$, which also equals to the maximum of $|f'(x)|$ on $[0, 1]$. We can assume $\alpha > 0$; otherwise, $f(x)$ would identically vanish. Suppose now that there is a point x_0 in $(0, 1)$ satisfying $g(x_0) > \alpha x_0$. By the mean value theorem there exists ξ in $(0, x_0)$ satisfying $g(x_0) = g'(\xi) x_0 > \alpha x_0$. However this implies $g'(\xi) > \alpha$, contrary to the definition of α. We thus have $g(x) \leq \alpha x$ for any $0 < x < 1$ so that $|f(x)| \leq \alpha \max\{x, 1-x\}$.

Hence

$$\int_0^1 |f(x)|\,dx \le \alpha \int_0^1 \max\{x, 1-x\}\,dx = \frac{\alpha}{4}.$$

The equality does not occur since the function $\max\{x, 1-x\}$ is not in C^1-class. However we can modify it slightly to become a continuously differentiable one in the neighborhood of $x = 1/2$ so that the difference between $\int_0^1 |f(x)|\,dx$ and $\alpha/4$ becomes sufficiently small.   □

SOLUTION 5.4

We introduce the auxiliary function

$$\phi(x) = \frac{x^n}{n+1}\int_0^x |f'(s)|^{n+1}\,ds - \int_0^x |f(s)|^n |f'(s)|\,ds.$$

Obviously $\phi(0) = 0$ and

$$\phi'(x) = \frac{nx^{n-1}}{n+1}\int_0^x |f'(s)|^{n+1}\,ds + \frac{x^n}{n+1}|f'(x)|^{n+1} - |f(x)|^n |f'(x)|.$$

Applying Hölder's inequality to 1 and $|f'(x)|$ we get

$$|f(x)| = \left|\int_0^x f'(s)\,ds\right| \le \int_0^x 1\cdot|f'(s)|\,ds$$
$$\le x^{n/(n+1)}\left(\int_0^x |f'(s)|^{n+1}\right)^{1/(n+1)},$$

or

$$\int_0^x |f'(s)|^{n+1}\,ds \ge \frac{|f(x)|^{n+1}}{x^n},$$

whence

$$\phi'(x) \ge \frac{n}{n+1}\cdot\frac{|f(x)|^{n+1}}{x} + \frac{x^n}{n+1}|f'(x)|^{n+1} - |f(x)|^n |f'(x)|.$$

The right-hand side multiplied by $(n+1)x$ is expressed as

$$\sigma(|f(x)|, x|f'(x)|),$$

where

$$\sigma(a, b) = na^{n+1} + b^{n+1} - (n+1)a^n b.$$

Integration 69

Since $\sigma(a, 0) \geq 0$, we can assume $b > 0$. Put $t = a/b \geq 0$ for brevity. Then

$$\frac{\sigma(a, b)}{b^{n+1}} = nt^{n+1} + 1 - (n + 1)t^n$$

attains its minimum 0 at $t = 1$, which implies that $\phi(x)$ is monotone increasing. In particular we have $\phi(1) \geq 0$, as required.

The equality occurs in Hölder's inequality if and only if $f(x)$ is linear. In this case the equality actually occurs in the inequality in question. ☐

SOLUTION 5.5

It suffices to show that

$$\int_0^1 f(x)\phi(x)\,dx \geq 0$$

where

$$\phi(x) = g(x) - \int_0^1 g(t)\,dt.$$

By the mean value theorem there is a ξ in $(0, 1)$ satisfying

$$g(\xi) = \int_0^1 g(x)\,dx.$$

Since $\phi(x) \leq 0$ for $0 \leq x \leq \xi$ and $\phi(x) \geq 0$ for $\xi \leq x \leq 1$, we have

$$\int_0^1 f(x)\phi(x)\,dx = \int_0^\xi f(x)\phi(x)\,dx + \int_\xi^1 f(x)\phi(x)\,dx$$

$$\geq f(\xi)\int_0^\xi \phi(x)\,dx + f(\xi)\int_\xi^1 \phi(x)\,dx$$

$$= f(\xi)\int_0^1 \phi(x)\,dx = 0.$$

☐

REMARK. Here is another proof due to Franklin (1885) using a double integral. Since

$$(f(x) - f(y))(g(x) - g(y))$$

is non-negative for any x and y in $[0, 1]$, it follows that

$$\iint_S (f(x) - f(y))(g(x) - g(y))\,dx\,dy \geq 0$$

where S is the unit square $[0, 1]^2$, whose expansion leads to the desired inequality.

SOLUTION 5.6

For a given polynomial

$$A(x) = x^n + a_1 x^{n-1} + \cdots + a_n$$

we define

$$B(x) = A\left(\frac{x+2}{4}\right), \qquad (5.4)$$

so that

$$B(x) = \frac{x^n}{4^n} + a'_1 x^{n-1} + \cdots + a'_n$$

with some real numbers $a'_1, \ldots, a'_n$. Putting further

$$Q(x) = \int_0^x B(s)\, ds \qquad (5.5)$$

we obtain

$$Q(x) = \frac{x^{n+1}}{4^n(n+1)} + a''_1 x^n + \cdots + a''_n x$$

with some real numbers $a''_1, \ldots, a''_n$. Applying the same method as in the proof of **PROBLEM 3.2** to $Q(x)$ with the same notations, we get

$$\frac{4(n+1)}{4^n(n+1)} = \left| \sum_{k=0}^{n+1}{}^{*}(-1)^k Q(\alpha_k) \right| \le \sum_{k=0}^{n} |Q(\alpha_k) - Q(\alpha_{k+1})| \qquad (5.6)$$

where

$$\alpha_k = 2\cos\frac{k\pi}{n+1}.$$

Therefore, by using (5.4) and (5.5) in (5.6),

$$\frac{1}{4^{n-1}} \le \sum_{k=0}^{n} \int_{\alpha_{k+1}}^{\alpha_k} |B(s)|\, ds = \int_{-2}^{2} |B(s)|\, ds = 4\int_0^1 |A(x)|\, dx.$$

$\square$

REMARK. The equality holds for

$$A_n(x) = \frac{1}{4^n(n+1)} T'_{n+1}(2x-1)$$

where $T_m(x)$ is the mth Chebyshev polynomial of the first kind. Indeed we have

$$
\int_0^1 |A_n(x)| \, dx = \frac{2}{4^{n+1}(n+1)} \int_{-1}^1 |T'_{n+1}(s)| \, ds
$$

$$
= \frac{2}{4^{n+1}(n+1)} \int_0^\pi |T'_{n+1}(\cos\theta)| \sin\theta \, d\theta
$$

$$
= \frac{2}{4^{n+1}} \int_0^\pi |\sin(n+1)\theta| \, d\theta = \frac{1}{4^n}.
$$

The polynomial

$$
U_n(x) = \frac{1}{n+1} T'_{n+1}(x)
$$

is called the nth Chebyshev polynomial of the second kind. $U_n(x)$ forms a system of orthogonal polynomials over $[-1, 1]$ and satisfies the relation

$$
U_n(\cos\theta) = \frac{\sin(n+1)\theta}{\sin\theta}.
$$

Achieser (1956) stated on p. 88 that this inequality is due to Korkin and Zolotareff (1873), while Chebyshev (1859) already got it implicitly. However Cheney (1966) stated on p. 233 that Korkin and Zolotareff (1873) posed the problem and it was Stieltjes (1876) who actually solved it.

SOLUTION 5.7

We have

$$
\sum_{k=1}^n f\left(\frac{k}{n}\right) - n \int_0^1 f(x) \, dx = n \sum_{k=1}^n \int_{(k-1)/n}^{k/n} \left(f\left(\frac{k}{n}\right) - f(x)\right) dx.
$$

By the mean value theorem there is a $\xi_{k,x}$ in each open interval $((k-1)/n, k/n)$ satisfying

$$
f\left(\frac{k}{n}\right) - f(x) = f'(\xi_{k,x})\left(\frac{k}{n} - x\right).
$$

Since $f'(x)$ is uniformly continuous on $[0, 1]$, we have for any $\epsilon > 0$

$$
\left| f'(\xi_{k,x}) - f'\left(\frac{k}{n}\right) \right| < \epsilon
$$

for all $1 \leq k \leq n$ and for all x in the interval $[(k-1)/n, k/n)$ on taking n sufficiently large. Hence

$$\left| S_n - \frac{1}{2n} \sum_{k=1}^{n} f'\left(\frac{k}{n}\right) \right| < \frac{\epsilon}{2}$$

where S_n is the expression in the problem. Since ϵ is arbitrary, we get

$$\lim_{n \to \infty} S_n = \frac{1}{2} \int_0^1 f'(x)\, dx = \frac{f(1) - f(0)}{2}.$$

$\square$

REMARK. One may think that this might be valid for any continuous function f on $[0, 1]$ since the expression does not contain the derivative of f. However we have the following counter-example.

Let $T(x)$ be the Takagi function in **PROBLEM 4.7**. Then we obtain

$$\sum_{k=1}^{2^n} T\left(\frac{k}{2^n}\right) - \sum_{k=1}^{2^n} \sum_{\ell=0}^{\infty} \frac{1}{2^\ell} g(2^{\ell-n}k) = \sum_{\ell=0}^{n-1} \frac{1}{2^\ell} \sum_{k=1}^{2^n} g(2^{\ell-n}k).$$

By the definition of g the right-hand side is equal to

$$\sum_{\ell=0}^{n-1} \left(2^{\ell-n+1} \sum_{k=1}^{2^{n-\ell-1}} k - \frac{1}{2} \right) = 2^{n-1} - \frac{1}{2}.$$

On the other hand,

$$\int_0^1 T(x)\, dx = \sum_{\ell=0}^{\infty} \frac{1}{2^\ell} \int_0^1 g(2^\ell x)\, dx = \frac{1}{4} \sum_{\ell=0}^{\infty} \frac{1}{2^\ell} = \frac{1}{2},$$

which implies that

$$S_{2^n} = -\frac{1}{2} \neq \frac{T(1) - T(0)}{2} = 0.$$

$\boxed{\text{SOLUTION } 5.8}$

(a) We show this by induction on n. When $n = 0$, (5.1) and (5.2) clearly hold for any $c_{1,0} \in (0, 1/2)$ and $\lambda_0 > 3^0 = 1$.

We next suppose that (5.1) holds for $n = m - 1$; that is,

$$\max_{0 \leq x \leq 1} \sum_{k=0}^{m-1} \psi_k(x) < 1 - \frac{1}{m+1}.$$

Then we can take a constant $c_{\ell,m}$ in the interval $(0, 1)$ satisfying

$$1 - \frac{1}{m+1} < c_{\ell,m} + \sum_{k=0}^{m-1} \psi_k(\alpha_{\ell,m}) < 1 - \frac{1}{m+2} \qquad (5.7)$$

for each $1 \leq \ell \leq 3^m$ and with $c_{\ell,m}$'s we define

$$\psi(x, \lambda) = \sum_{\ell=1}^{3^m} c_{\ell,m} \phi\big(\lambda(x - \alpha_{\ell,m})\big).$$

Since

$$\lim_{\lambda \to \infty} \psi(x, \lambda) = \begin{cases} c_{\ell,m} & \text{if} \quad x = \alpha_{\ell,m}, \\ 0 & \text{otherwise,} \end{cases}$$

we can take a sufficiently large $\lambda_m > 3^{5m+1}$ such that $\psi(\alpha_{\ell,m}, \lambda_m) = \psi_k(\alpha_{\ell,m})$ is sufficiently close to $c_{\ell,m}$ for all $1 \leq \ell \leq 3^m$. Substituting this in (5.7), we conclude that

$$1 - \frac{1}{m+1} < \sum_{k=0}^{m} \psi_k(\alpha_{\ell,m}) < 1 - \frac{1}{m+2}$$

for any $1 \leq \ell \leq 3^m$ and that

$$\max_{0 \leq x \leq 1} \sum_{k=0}^{m} \psi_k(x) < 1 - \frac{1}{m+2}.$$

This shows that (5.1) and (5.2) hold for $n = m$.

By the property (a) we see that the partial sums of

$$\Phi(x) = \sum_{n=0}^{\infty} \psi_n(x)$$

is a monotone increasing sequence and so the series converges pointwise. Moreover $0 < \Phi(x) \leq 1$ for any $x \in [0, 1]$ and $\Phi(x) = 1$ for any $x \in A$, where

$$A = \big\{ \alpha_{\ell,m} ; \ 1 \leq \ell \leq 3^m, \ m \geq 0 \big\}$$

is a dense subset of the interval $[0, 1]$.

To see that $\Phi(x)$ is not constant on any subinterval of $[0, 1]$, we put $\beta_{k,n} = k3^{-n}$ for $0 \leq k \leq 3^n$ and

$$B = \big\{ \beta_{k,n} ; \ 0 \leq \ell \leq 3^n, \ n \geq 0 \big\}.$$

Problems and Solutions in Real Analysis

B is also a dense subset of $[0, 1]$ satisfying $A \cap B = \emptyset$. Since

$$\left| \alpha_{j,m} - \beta_{k,n} \right| = \left| \frac{2j-1}{2 \cdot 3^m} - \frac{k}{3^n} \right| \geq \frac{1}{2 \cdot 3^m}$$

for any integers j, k and $m \geq n$, we have

$$\psi_m(\beta_{k,n}) = \sum_{j=1}^{3^m} c_{j,m} \phi\left(\lambda_m \left(\beta_{k,n} - \alpha_{j,m} \right) \right)$$

$$< 3^m \sqrt{\frac{2 \cdot 3^m}{\lambda_m}} < \frac{1}{3^m}.$$

Therefore

$$\Phi(\beta_{k,n}) = \sum_{m=0}^{n-1} \psi_m(\beta_{k,n}) + \sum_{m=n}^{\infty} \psi_m(\beta_{k,n})$$

$$< 1 - \frac{1}{n+1} + \sum_{m=n}^{\infty} \frac{1}{3^m},$$

which is less than 1 for n large enough. We thus have $\Phi(x) < 1$ for any $x \in B$.

(b) Let V be the set of all positive continuous functions f defined on $\mathbb{R}$ satisfying $\sigma_{a,b}(f) < 4 \min\{ f(a), f(b) \}$ for any $a \neq b$, where

$$\sigma_{a,b}(f) = \frac{1}{b-a} \int_a^b f(x) \, dx.$$

We first show that $\phi \in V$. Since ϕ is an even function, it suffices to consider only two cases: $0 \leq a < b$ and $a < 0 < b$. If $0 \leq a < b$, then

$$\sigma_{a,b}(\phi) = \frac{2}{\sqrt{1+a} + \sqrt{1+b}} < 2\phi(b).$$

If $a < 0 < b$, then

$$\sigma_{a,b}(\phi) = \frac{2}{|a|+b} \left(\sqrt{1+|a|} + \sqrt{1+b} - 2 \right),$$

which is less than

$$\frac{4}{c} \left(\sqrt{1+c} - 1 \right) = \frac{4}{\sqrt{1+c} + 1} < 4\phi(c),$$

where $c = \max\{|a|, b\}$, as required.

Note that V forms a positive cone; that is, if f_1 and f_2 belong to V, then $c_1 f_1 + c_2 f_2$ also belongs to V for any positive constants c_1 and c_2. Moreover, for any $f \in V$ and $\lambda > 0$, the function $\tilde{f}$ defined by $\tilde{f}(x) = f(\lambda x)$ belongs to V in view of

$$\sigma_{a,b}(\tilde{f}) = \sigma_{\lambda a, \lambda b}(f) < 4 \min\{f(\lambda a), f(\lambda b)\} = 4 \min\{\tilde{f}(a), \tilde{f}(b)\}.$$

This means that every $\psi_k(x)$ defined above belongs to V.

Hence, noting that

$$\left| \int_0^x \psi_n(t)\, dt \right| \leq 4\psi_n(0),$$

we infer that the series $\Psi(x)$ given in (5.3) converges absolutely and uniformly for $0 \leq x \leq 1$. Therefore $\Psi(x)$ is continuous on the interval $[0, 1]$. Moreover, for an arbitrary fixed $x \in (0, 1)$ and any $\epsilon > 0$, we can take an integer $N = N(x, \epsilon)$ satisfying

$$\sum_{n=N}^{\infty} \psi_n(x) < \epsilon.$$

We then take a sufficiently small number $\delta > 0$ such that

$$|\psi_k(\xi) - \psi_k(x)| < \frac{\epsilon}{N}$$

for any integer $0 \leq k < N$ and any ξ with $|x - \xi| < \delta$. Thus, for any $0 < |h| < \delta$, we obtain

$$\left| \frac{\Psi(x+h) - \Psi(x)}{h} - \sum_{n=0}^{\infty} \psi_n(x) \right| = \left| \sum_{n=0}^{\infty} \frac{1}{h} \int_x^{x+h} (\psi_n(t) - \psi_n(x))\, dt \right|$$

$$\leq \epsilon + 2 \sum_{n=N}^{\infty} \psi_n(x) < 3\epsilon,$$

which implies that $\Psi(x)$ is differentiable and satisfies

$$\Psi'(x) = \sum_{n=0}^{\infty} \psi_n(x)$$

for any $x \in (0, 1)$.

(c) Finally we construct an example of everywhere differentiable but nowhere monotone functions. Let

$$f(x) = \Psi(x) - \Psi\left(x - \frac{1}{6}\right)$$

for $1/6 < x < 1$. Since $\alpha_{j,n} - 1/6 = \beta_{k,n}$ for any $k = j - (3^{n-1} + 1)/2 \geq 0$, we have

$$f'(\alpha_{j,n}) = \Phi(\alpha_{j,n}) - \Phi(\beta_{k,n}) > 0.$$

Similarly, since $\beta_{k,n} - 1/6 = \alpha_{j,n}$ for any $j = k - (3^{n-1} - 1)/2 \geq 1$, we get

$$f'(\beta_{k,n}) = \Phi(\beta_{k,n}) - \Phi(\alpha_{j,n}) < 0.$$

Thus, A and B being dense subsets of $[0, 1]$, $f(x)$ is nowhere monotone. $\square$

$\boxed{\text{SOLUTION 5.9}}$

As is already seen in **SOLUTION 1.7**, the maximal values in the interval $[0, \pi]$ of the function

$$S_m(\theta) = \sin\theta + \frac{\sin 2\theta}{2} + \cdots + \frac{\sin m\theta}{m}$$

are attained at $[(m + 1)/2]$ points: $\pi/(m + 1)$, $3\pi/(m + 1)$, ... Since

$$S'_m(\theta) = \frac{1}{2}\sin 2(m + 1)\vartheta \cot\vartheta - \cos^2(m + 1)\vartheta,$$

we get

$$S_m(\beta) - S_m(\alpha) \leq \int_{\alpha/2}^{\beta/2} \sin 2(m + 1)\vartheta \cot\vartheta \, d\vartheta$$

for any $0 < \alpha < \beta < \pi$. By the substitution $s = 2(m + 1)\vartheta - 2\ell\pi$ with

$$\alpha = \frac{2\ell - 1}{m + 1}\pi \quad \text{and} \quad \beta = \frac{2\ell + 1}{m + 1}\pi,$$

the above integral on the right-hand side can be written as

$$\frac{1}{2(m + 1)}\int_0^\pi \sin s\left(\cot\frac{2\ell\pi + s}{2(m + 1)} - \cot\frac{2\ell\pi - s}{2(m + 1)}\right)ds.$$

Since the function $\cot s$ is strictly monotone decreasing in the interval $(0, \pi/2)$, this integral is clearly negative. This implies that the maximum of $S_m(\theta)$ on the interval $[0, \pi]$ is attained at $\theta = \pi/(m + 1)$. Moreover we have

$$S_{m+1}\left(\frac{\pi}{m + 2}\right) > S_{m+1}\left(\frac{\pi}{m + 1}\right) = S_m\left(\frac{\pi}{m + 1}\right).$$

Therefore $\{S_m(\pi/(m + 1))\}$ is a strictly monotone increasing sequence and

$$S_m\left(\frac{\pi}{m + 1}\right) = \frac{\pi}{m + 1}\sum_{n=1}^{m+1}\frac{m + 1}{n\pi}\sin\frac{n\pi}{m + 1}$$

converges to the integral $\int_0^\pi \frac{\sin x}{x}\,dx$ as $m \to \infty$. $\square$

Chapter 6

Improper Integrals

In the previous chapter the integral was defined for bounded functions on closed bounded intervals. The notion of integrals can be extended to (unbounded) functions defined on open (unbounded) intervals.

- Suppose that a function $f(x)$ defined on $[a, b)$ is integrable in the sense of Riemann on the interval $[a, b - \epsilon]$ for any ϵ in $(0, b - a)$. If the Riemann integral

$$\int_a^{b-\epsilon} f(x)\, dx$$

  converges as $\epsilon \to 0+$, then the limit is denoted by $\int_a^b f(x)\, dx$ and called the *improper Riemann integral* of $f(x)$ over $[a, b)$. The improper integrals can similarly be defined for bounded intervals like $(a, b]$ or (a, b).

- If a function $f(x)$ defined on the interval $[a, \infty)$ is integrable in the sense of Riemann on $[a, b]$ for any $b > a$, we define $\int_a^\infty f(x)\, dx$ as the limit of $\int_a^b f(x)\, dx$ as $b \to \infty$ if it exists. Similarly we can define the improper integrals for unbounded intervals like $(-\infty, b]$ or $(-\infty, \infty)$.

The convergence of the improper integral of $f(x)$ does not imply that of $|f(x)|$ in general. For example,

$$\int_0^\infty \frac{\sin x}{\sqrt{x}}\, dx = \sqrt{\pi/2}\,,$$

known as Fresnel's integral, while $\int_0^\infty |\sin x| / \sqrt{x}\, dx$ diverges.

- If a continuous function $f(x)$ defined on $[1, \infty)$ is positive and monotone decreasing, then $\sum_{n=1}^{\infty} f(n)$ converges if and only if $\int_{1}^{\infty} f(x)\,dx$ converges.

PROBLEM 6.1

Show that

$$\int_{0}^{1} x^{-x}\,dx = \frac{1}{1^1} + \frac{1}{2^2} + \frac{1}{3^3} + \cdots + \frac{1}{n^n} + \cdots .$$

Although no use of trick is to be made in the proof, the expression is interesting in the sense that

$$\int_{0}^{1} f(x)\,dx = \sum_{n=1}^{\infty} f(n)$$

holds for $f(x) = x^{-x}$. We also have another example $f(x) = a^{-x}$ where a is a unique real root greater than 1 of the equation

$$a - 2 + \frac{1}{a} = \log a.$$

It is remarkable that

$$\int_{0}^{1} x^{x}\,dx = \frac{1}{1^1} - \frac{1}{2^2} + \frac{1}{3^3} - \frac{1}{4^4} + \cdots$$

was already noticed by Johannis Bernoulli (1697). This integral appeared also in the William Lowell Putnam Mathematical Competition (1970) and in Elementary Problems and Solutions proposed by Klamkin (1970) in 'American Mathematical Monthly'. See also p. 308 of Ramanujan's Notebooks Part IV edited by Berndt (1994).

PROBLEM 6.2

Show that

$$\lim_{n \to \infty} \sqrt{n} \int_{-\infty}^{\infty} \frac{dx}{(1 + x^2)^n} = \sqrt{\pi}.$$

Improper Integrals 79

PROBLEM 6.3

Show that

$$\int_0^\infty \frac{e^{-x/s-1/x}}{x}\,dx \sim \log s$$

as $s \to \infty$.

PROBLEM 6.4

Suppose that $g \in C[0,\infty)$ *is monotone decreasing and that* $\int_0^\infty g(x)\,dx$ *converges.* (*Note that* $g(x) \geq 0$ *for any* $x \geq 0$.) *Show then that*

$$\lim_{h\to 0+} h \sum_{n=1}^\infty f(nh) = \int_0^\infty f(x)\,dx$$

for any $f \in C[0,\infty)$ *satisfying* $|f(x)| \leq g(x)$ *for all* $x \geq 0$.

PROBLEM 6.5

For $s > 0$ *compute*

$$\int_0^\infty e^{-(x-s/x)^2}\,dx.$$

PROBLEM 6.6

Suppose that $f \in C(\mathbb{R})$ *and that* $\int_{-\infty}^\infty |f(x)|\,dx$ *converges. Show then that*

$$\lim_{n\to\infty} \int_{-\infty}^\infty f(x)|\sin nx|\,dx = \frac{2}{\pi}\int_{-\infty}^\infty f(x)\,dx.$$

PROBLEM 6.7

Show that

$$\frac{7}{12} - \gamma = \int_0^\infty \frac{\{x\}^2\,(1-\{x\})^2}{(1+x)^5}\,dx$$

where γ *is Euler's constant and* $\{x\}$ *denotes the fractional part of* x.

PROBLEM 6.8

Show that a rational function $R(x)$ satisfies

$$\int_{-\infty}^{\infty} f(R(x))\,dx = \int_{-\infty}^{\infty} f(x)\,dx$$

for all piecewise continuous function $f(x)$ such that $\int_{-\infty}^{\infty} f(x)\,dx$ exists, if and only if

$$R(x) = \pm\left(x - \alpha_0 - \sum_{k=1}^{m} \frac{c_k}{x - \alpha_k} \right)$$

for some non-negative integer m, real constants $\alpha_0, \alpha_1, \ldots, \alpha_m$ with $\alpha_1 < \cdots < \alpha_m$ and positive constants $c_1, \ldots, c_m$.

This is the problem posed by Pólya (1931) and solved by Szegö (1934).

PROBLEM 6.9

Show that

$$\gamma = \int_0^1 \frac{1 - \cos x}{x}\,dx - \int_1^{\infty} \frac{\cos x}{x}\,dx,$$

where γ is Euler's constant.

This formula will be used in the proof of Kummer's series for $\log \Gamma(s)$. See **PROBLEM 16.11**. It can be easily verified by differentiation that

$$\log s + \gamma = \int_0^s \frac{1 - \cos x}{x}\,dx - \int_s^{\infty} \frac{\cos x}{x}\,dx$$

for any $s > 0$.

Solutions for Chapter 6

SOLUTION 6.1

We have

$$\int_0^1 x^{-x}\,dx = \int_0^1 e^{-x\log x}\,dx = \sum_{n=0}^{\infty} \frac{(-1)^n}{n!} \int_0^1 x^n \log^n x\,dx,$$

where the termwise integration is allowed since the series

$$\sum_{n=0}^{\infty} \frac{(-1)^n}{n!} x^n \log^n x$$

converges uniformly on the interval $(0, 1]$. Also by the substitution $x = e^{-s}$ we get

$$\int_0^1 x^n \log^n x\,dx = (-1)^n \int_0^{\infty} e^{-(n+1)s} s^n\,ds = (-1)^n \frac{n!}{n^n}$$

from the definition of the Gamma function (See Chapter 16). $\qquad\square$

SOLUTION 6.2

We divide $(-\infty, \infty)$ into three parts as

$$\left(-\infty, -n^{-1/3}\right) \cup \left[-n^{-1/3}, n^{-1/3}\right] \cup \left(n^{-1/3}, \infty\right),$$

which we denote by A_1, A_2, A_3 respectively. For $k = 1, 2, 3$ put

$$I_k = \sqrt{n} \int_{A_k} \frac{dx}{(1 + x^2)^n}.$$

By the substitution $t = \sqrt{n}\,x$ we have

$$I_2 = \sqrt{n} \int_{A_2} \exp\left(-n\log\left(1 + x^2\right)\right) dx$$

$$= \int_{-n^{1/6}}^{n^{1/6}} \exp\left(-n\log\left(1 + \frac{t^2}{n}\right)\right) dt.$$

Since

$$n \log\left(1 + \frac{t^2}{n}\right) = t^2 + O\left(\frac{1}{n^{1/3}}\right)$$

uniformly in $|t| \le n^{1/6}$, we obtain

$$I_2 = \left(1 + O\left(\frac{1}{n^{1/3}}\right)\right)\left(\sqrt{\pi} - \int_{-\infty}^{-n^{1/6}} e^{-t^2} \, dt - \int_{n^{1/6}}^{\infty} e^{-t^2} \, dt\right)$$

$$= \sqrt{\pi} + O\left(\frac{1}{n^{1/3}}\right)$$

as $n \to \infty$. On the other hand it follows that

$$0 < I_1 + I_3 < \frac{\sqrt{n}}{(1 + n^{-2/3})^{n-1}} \int_{-\infty}^{\infty} \frac{dx}{1 + x^2}$$

and the right-hand side converges to 0 as $n \to \infty$. □

Solution 6.3

Let $I(s)$ be the improper integral in the problem, which is invariant under the substitution $t = s/x$; hence

$$I(s) = 2 \int_{\sqrt{s}}^{\infty} \frac{e^{-x/s - 1/x}}{x} \, dx.$$

Since $e^{-1/x} = 1 + O(s^{-1/2})$ uniformly in $x \ge \sqrt{s}$ as $s \to \infty$, we have

$$I(s) = 2\left(1 + O\left(\frac{1}{\sqrt{s}}\right)\right) \int_{\sqrt{s}}^{\infty} \frac{e^{-x/s}}{x} \, dx$$

$$= 2\left(1 + O\left(\frac{1}{\sqrt{s}}\right)\right) \int_{s^{-1/2}}^{\infty} \frac{e^{-t}}{t} \, dt.$$

Integrating the last integral by parts, we get

$$\int_{s^{-1/2}}^{\infty} \frac{e^{-t}}{t} \, dt = \left[e^{-t} \log t\right]_{t=s^{-1/2}}^{t=\infty} + \int_{s^{-1/2}}^{\infty} e^{-t} \log t \, dt$$

$$= \frac{1}{2} \log s + O(1);$$

hence $I(s) = \log s + O(1)$ as $s \to \infty$. □

SOLUTION 6.4

For any $\epsilon > 0$ there exists a sufficiently large number $L > 1$ satisfying

$$\int_{L-1}^{\infty} g(x)\,dx < \epsilon.$$

For any h in the interval $(0, 1)$ take a positive integer N satisfying

$$Nh \leq L < (N+1)h.$$

Then

$$nh \in \left(\frac{Ln}{N+1}, \frac{Ln}{N}\right] \subset \left(\frac{L(n-1)}{N}, \frac{Ln}{N}\right]$$

for any integer $1 \leq n \leq N$. Thus the Riemann sum

$$\frac{L}{N} \sum_{n=1}^{N} f(nh)$$

converges to the integral $\int_{0}^{L} f(x)\,dx$ as $N \to \infty$. Since $N \to \infty$ as $h \to 0+$, there exists a sufficiently small $h_0 > 0$ such that

$$\left| \frac{L}{N} \sum_{n=1}^{N} f(nh) - \int_{0}^{L} f(x)\,dx \right| < \epsilon \tag{6.1}$$

and

$$\frac{1}{N} \int_{0}^{\infty} g(x)\,dx < \epsilon$$

for any $0 < h < h_0$. On the other hand, we have

$$\left| \left(\frac{L}{N} - h\right) \sum_{n=1}^{N} f(nh) \right| \leq \left| \frac{L}{N} - h \right| \sum_{n=1}^{N} g(nh) \leq \frac{h}{N} \sum_{n=1}^{N} g(nh).$$

By the monotonicity of $g(x)$ the right-hand side is less than or equal to

$$\frac{1}{N} \int_{0}^{L} g(x)\,dx,$$

which is clearly less than ϵ, whence

$$\left| \left(\frac{L}{N} - h\right) \sum_{n=1}^{N} f(nh) \right| < \epsilon. \tag{6.2}$$

Moreover

$$h \left| \sum_{n > N} f(nh) \right| \leq h \sum_{n > N} g(nh) \leq \int_{L-1}^{\infty} g(x) \, dx < \epsilon; \qquad (6.3)$$

therefore, by (6.1), (6.2) and (6.3),

$$\left| h \sum_{n=1}^{\infty} f(nh) - \int_{0}^{\infty} f(x) \, dx \right| < 3\epsilon + \int_{L}^{\infty} |f(x)| \, dx < 4\epsilon.$$

Since ϵ is arbitrary, this completes the proof. □

REMARK. The special case $f(x) = g(x)$ was shown in Pólya and Szegö (1972).

SOLUTION 6.5

Let $f(s)$ be the integral in the problem. It is easily seen that

$$e^{-4s} f(s) = \int_{0}^{\infty} \exp\left(-(x + s/x)^2\right) dx.$$

Differentiating both sides we get

$$(f'(s) - 4f(s)) e^{-4s} = -2 \int_{0}^{\infty} \left(1 + \frac{s}{x^2}\right) \exp\left(-(x + s/x)^2\right) dx,$$

where the differentiation under the integral sign is allowed (See the introduction of Chapter 11). Then, by the substitution $t = x - s/x$, we have

$$f'(s) - 4f(s) = -2 \int_{0}^{\infty} \left(1 + \frac{s}{x^2}\right) \exp\left(-(x + s/x)^2\right) dx$$

$$= -2 \int_{-\infty}^{\infty} e^{-t^2} \, dt = -2\sqrt{\pi}.$$

Solving this linear differential equation we get

$$f(s) = c e^{4s} + \frac{\sqrt{\pi}}{2}$$

for some constant c. Since $0 < f(s) < (\sqrt{\pi}/2) e^{2s}$, we have $c = 0$; hence

$$f(s) = \frac{\sqrt{\pi}}{2}.$$

□

REMARK. This may be also solved by applying **PROBLEM 6.8**. Note that the function $f(s)$ can be defined for all real numbers s and $f(s) = \sqrt{\pi}/2$ is valid also for $s = 0$. The above method is effective even for negative s and we will get the differential equation

$$f'(s) = 4f(s),$$

from which we have

$$f(s) = \frac{\sqrt{\pi}}{2} e^{4s}$$

for negative s.

$\boxed{\text{SOLUTION 6.6}}$

For any $\epsilon > 0$ we can take a sufficiently large number L satisfying

$$\int_{-\infty}^{-L\pi} + \int_{L\pi}^{\infty} |f(x)| \, dx < \epsilon. \tag{6.4}$$

On the other hand,

$$\int_{-L\pi}^{L\pi} f(x) |\sin nx| \, dx = \pi \int_{-L}^{L} f(\pi t) |\sin n\pi t| \, dt$$

$$= \pi \sum_{k=-L}^{L-1} \int_{0}^{1} f(\pi s + k\pi) |\sin n\pi s| \, ds.$$

Applying the result in **PROBLEM 5.1**, the right-hand side converges to

$$\pi \sum_{k=-L}^{L-1} \int_{0}^{1} f(\pi s + k\pi) \, ds \int_{0}^{1} \sin \pi s \, ds$$

as $n \to \infty$, which is

$$\frac{2}{\pi} \int_{-L\pi}^{L\pi} f(t) \, dt,$$

whence

$$\int_{-L\pi}^{L\pi} f(x) |\sin nx| \, dx \to \frac{2}{\pi} \int_{-L\pi}^{L\pi} f(t) \, dt \quad (n \to \infty). \tag{6.5}$$

Therefore , by (6.4) and (6.5), we have

$$\limsup_{n \to \infty} \left| \int_{-\infty}^{\infty} f(x) |\sin nx| \, dx - \frac{2}{\pi} \int_{-\infty}^{\infty} f(t) \, dt \right| \leq 2\epsilon.$$

Since ϵ is arbitrary, this completes the proof. $\qquad\square$

SOLUTION 6.7

Let I be the integral on the right-hand side of the problem. We have

$$
I = \sum_{k=1}^{\infty} \int_{k-1}^{k} \frac{\{x\}^2 (1 - \{x\})^2}{(1 + x)^5} \, dx
$$
$$
= \int_0^1 t^2 (1 - t)^2 H_5(t) \, dt,
$$

where for any integer $m > 1$ we write

$$
H_m(x) = \sum_{k=1}^{\infty} \frac{1}{(x + k)^m}.
$$

Then, by integration by parts, we get

$$
I = \int_0^1 t(1 - t)\left(\frac{1}{2} - t\right) H_4(t) \, dt
$$
$$
= \int_0^1 \left(\frac{1}{6} - t(1 - t)\right) H_3(t) \, dt.
$$

Since

$$
\int_0^1 H_3(t) \, dt = -\frac{1}{2} \left(H_2(1) - H_2(0)\right) = \frac{1}{2},
$$

it follows that

$$
I = \frac{1}{12} + \int_0^1 \left(t - \frac{1}{2}\right) H_2(t) \, dt \,;
$$

therefore

$$
I = \frac{1}{12} + \lim_{n \to \infty} \sum_{k=1}^{n} \int_0^1 \frac{t - 1/2}{(t + k)^2} \, dt
$$
$$
= \frac{1}{12} - \lim_{n \to \infty} \sum_{k=1}^{n} \left(\frac{1}{2k} + \frac{1}{2(k + 1)} - \log \frac{k + 1}{k}\right),
$$

which is equal to $7/12 - \gamma$ from the definition of Euler's constant. $\qquad\square$

SOLUTION 6.8

The proof is based on Szegö (1934).

For any ϵ in the interval $(0, 1)$ we can find a positive constant M satisfying

$$|R(x) - R(x_0)| \leq |R'(x_0)|\epsilon + M\epsilon^2$$

for any $|x - x_0| \leq \epsilon$ unless x_0 is a real pole of $R(x)$. Let $f_\epsilon(x)$ be the piecewise continuous function defined by

$$f_\epsilon(x) = \begin{cases} 1 & \text{if} \quad |x - R(x_0)| \leq |R'(x_0)|\epsilon + M\epsilon^2, \\ 0 & \text{otherwise.} \end{cases}$$

Then we have

$$2\epsilon(|R'(x_0)| + M\epsilon) = \int_{-\infty}^{\infty} f_\epsilon(x)\, dx$$

$$\geq \int_{x_0-\epsilon}^{x_0+\epsilon} f_\epsilon(R(x))\, dx = 2\epsilon;$$

hence, ϵ being arbitrary, we have $|R'(x_0)| \geq 1$. This means that $R(x)$ maps each interval divided by real poles of $R(x)$, say $\alpha_1 < \alpha_2 < \cdots < \alpha_m$, bijectively on $\mathbb{R}$. Of course $m = 0$ corresponds to the case in which $R(x)$ has no real poles. The equation $R(x) = s$ has a unique solution in each interval $(-\infty, \alpha_1)$, (α_1, α_2), ..., (α_m, ∞), say $x_k(s)$, $0 \leq k \leq m$.

Next let $g_\epsilon(x)$ be the piecewise continuous function defined by

$$g_\epsilon(x) = \begin{cases} 1 & \text{if} \quad |x - s| \leq \epsilon, \\ 0 & \text{otherwise.} \end{cases}$$

We then have

$$2\epsilon = \int_{-\infty}^{\infty} g_\epsilon(x)\, dx = \int_{-\infty}^{\infty} g_\epsilon(R(x))\, dx, \tag{6.6}$$

where the right-hand side is equal to the sum of lengths of $m+1$ intervals satisfying $|R(x) - s| \leq \epsilon$. Each of these intervals contains exactly one $x_k(s)$ and the length is

$$\frac{2\epsilon}{|R'(x_k(s))|} + O(\epsilon^2);$$

so letting ϵ to $0+$ in (6.6) we get

$$\sum_{k=0}^{m} \frac{1}{|R'(x_k(s))|} = 1. \tag{6.7}$$

In the case $m = 0$ the function $R(x)$ maps $\mathbb{R}$ homeomorphically onto $\mathbb{R}$ and $R'(x_0(s)) = \pm 1$ implies that

$$x_0'(s) = (R^{-1})'(s) = \frac{1}{R'(R^{-1}(s))} = \frac{1}{R'(x_0(s))} = \pm 1.$$

Therefore $R(x) = \pm(x - \alpha_0)$ for some constant α_0 for $m = 0$.

We hereafter assume that m is a positive integer. Since $|R'(x_k(s))|$ diverges to ∞ as $|s| \to \infty$ for each $1 \leq k < m$, it follows from (6.7) that either

$$\begin{cases} R'(x) \geq 1 \quad \text{on} \quad (-\infty, \alpha_1) \cup (\alpha_m, \infty), \\[2mm] \lim_{s \to -\infty} x_0(s) = -\infty, \quad \lim_{s \to -\infty} x_m(s) = \alpha_m, \\[2mm] \lim_{s \to \infty} x_0(s) = \alpha_1, \quad \lim_{s \to \infty} x_m(s) = \infty \end{cases}$$

or

$$\begin{cases} R'(x) \leq -1 \quad \text{on} \quad (-\infty, \alpha_1) \cup (\alpha_m, \infty), \\[2mm] \lim_{s \to -\infty} x_0(s) = \alpha_1, \quad \lim_{s \to -\infty} x_m(s) = \infty, \\[2mm] \lim_{s \to \infty} x_0(s) = -\infty, \quad \lim_{s \to \infty} x_m(s) = \alpha_m \end{cases}$$

holds. They are called the first and the second cases respectively.

The rational function $R(x)$ can now be represented as

$$R(x) = P(x) - \sum_{k,\ell} \frac{c_{k,\ell}}{(x - \alpha_k)^{d_{k,\ell}}} + Q(x)$$

where $P(x)$ is a polynomial of degree r with some positive integer r, $c_{k,\ell}$ are non-zero real constants, $d_{k,\ell}$ are positive integers, and $Q(x)$ is a rational function having no real poles and converging to 0 as $|x| \to \infty$ unless $Q(x)$ vanishes identically. Let βx^r be the leading term of $P(x)$ where β is a non-zero constant. Since $R'(x) \sim \beta r x^{r-1}$ as $|x| \to \infty$, we must have $r = 1$ and hence $\beta = \pm 1$ by (6.7). Hence $P(x) = \pm(x - \alpha_0)$ for some constant α_0.

We first treat the first case. Since $R'(x) = 1 + O(x^{-2})$ as $|x| \to \infty$ in this case, we have

$$\begin{aligned} \sum_{k=0}^{m-1} \frac{1}{|R'(x_k(s))|} &= O(s^{-2}) \quad \text{as} \quad s \to \infty, \\[2mm] \sum_{k=1}^{m} \frac{1}{|R'(x_k(s))|} &= O(s^{-2}) \quad \text{as} \quad s \to -\infty. \end{aligned} \tag{6.8}$$

Let d_k^* be the largest integer among $d_{k,\ell}$ and c_k^* be the corresponding coefficient $c_{k,\ell}$ for each $0 \le k \le m$. Clearly $x_k(s) \to \alpha_k$ either as $s \to \infty$ or $s \to -\infty$ for $1 \le k \le m$. Then we have

$$R(x) \sim -\frac{c_k^*}{(x - \alpha_k)^{d_k^*}} \quad \text{as} \quad x \to \alpha_k,$$

and in particular,

$$s = R(x_k(s)) \sim -\frac{c_k^*}{(x_k(s) - \alpha_k)^{d_k^*}},$$

whence solving in $x_k(s) - \alpha_k$ and substituting in

$$R'(x_k(s)) \sim \frac{c_k^* d_k^*}{(x_k(s) - \alpha_k)^{d_k^*+1}} \quad \text{as} \quad x \to \alpha_k,$$

we obtain

$$\frac{1}{|R'(x_k(s))|} \sim \frac{|c_k^*|^{1/d_k^*}}{d_k^*} \cdot \frac{1}{|s|^{1+1/d_k^*}}$$

either as $s \to \infty$ or $s \to -\infty$. Therefore, in view of (6.8), $d_k^* = 1$ and $R(x)$ can be written as

$$R(x) = x - \alpha_0 - \sum_{k=1}^{m} \frac{c_k}{x - \alpha_k} + Q(x) \tag{6.9}$$

for some real constants c_k in the first case. Since $R(x)$ is monotone increasing on (α_m, ∞) we have $c_m > 0$ and hence all coefficients c_k must be positive; in particular,

$$\sum_{k=0}^{m} \frac{1}{R'(x_k(s))} = 1. \tag{6.10}$$

Differentiating $s = R(x_k(s))$ in s and substituting in (6.10), we get

$$\sum_{k=0}^{m} x_k'(s) = 1,$$

from which it follows that

$$\sum_{k=0}^{m} x_k(s) = s + c'$$

for some constant c'. Since $x_m(s) = s + \alpha_0 + O(s^{-1})$ and $x_k(s) = \alpha_{k+1} + O(s^{-1})$
as $s \to \infty$ for each $0 \le k < m$, we have

$$\sum_{k=0}^{m} x_k(s) = s + \sum_{k=0}^{m} \alpha_k. \tag{6.11}$$

The rest of the proof is devoted to showing the vanishing of Q. It may be interesting to find an easier real-analytic proof of this part.

Suppose, on the contrary, that $Q(x) \not\equiv 0$. Then, in view of (6.9), we can write $R(x) = V(x)/U(x)$ with

$$U(x) = A(x) \prod_{k=1}^{m} (x - \alpha_k)$$

where

$$\begin{cases} A(x) = x^{2p} - c'' x^{2p-1} + O(x^{2p-2}), \\[2mm] V(x) = x^{m+2p+1} - \left(c'' + \sum_{k=0}^{m} \alpha_k \right) x^{m+2p} + O(x^{m+2p-1}) \end{cases}$$

are polynomials with real coefficients for some positive integer p and some constant c''. We can assume that $U(x)$ and $V(x)$ are relatively prime; that is, they have no common factor except for constants. Let $w_1, \overline{w_1}, \ldots, w_p, \overline{w_p}$ be non-real zeros of $A(z)$. The algebraic equation

$$V(z) = sU(z)$$

has exactly $m + 1$ real simple roots $x_0(s), \ldots, x_m(s)$ and $2p$ non-real roots $z_1(s)$, $\overline{z_1(s)}, \ldots, z_p(s), \overline{z_p(s)}$ counting with multiplicity for any real number s. Then we have from (6.11)

$$z_1(s) + \overline{z_1(s)} + \cdots + z_p(s) + \overline{z_p(s)} = c''. \tag{6.12}$$

Let C be a circle enclosing all the zeros of $U(z)$. Take a sufficiently large s such that

$$s \min_{z \in C} |U(z)| > \max_{z \in C} |V(z)|$$

and that $x_m(s)$ lies outside of C. Since $|V(z)| < s|U(z)|$ on C, it follows from Rouché's theorem that $V(z) = sU(z)$ has exactly $m + 2p$ roots inside of C, which are of course $x_1(s), \ldots, x_m(s)$ and $z_1(s), \overline{z_1(s)}, \ldots, z_p(s), \overline{z_p(s)}$. This implies that all the non-real roots are bounded as $s \to \infty$.

$V(z) - sU(z)$ is irreducible as a polynomial of two variables; that is, it cannot be expressed as the product of two polynomials none of which is constant. We then consider $x_k(s)$ and $z_k(s)$ as function elements of the algebraic function uniquely determined by $V(z) - sU(z) = 0$. Since all solutions of $V(z) - sU(z) = 0$ are branches of the same algebraic function, it follows in particular that $x_m(s)$ can be continued to $z_1(s)$ along an arc on the Riemann surface of this algebraic function. This is a contradiction, since the relation (6.12) holds globally except for possible isolated singularities and since $x_m(s)$ is a unique solution which is unbounded as $s \to \infty$. Therefore $p = 0$ and hence $Q(x)$ must vanish identically, completing the proof of the first part.

The similar argument can be applied to the second case.

Conversely let

$$R(x) = \pm \left(x - \alpha_0 - \sum_{k=1}^{m} \frac{c_k}{x - \alpha_k} \right)$$

for some real numbers $\alpha_0, ..., \alpha_m$ with $\alpha_1 < \cdots < \alpha_m$ and some positive constants $c_1, ..., c_m$. Then it is clear that $\sum_{k=0}^{m} x'_k(s) = \pm 1$ and hence

$$\int_{-\infty}^{\infty} f(R(x)) \, dx = \int_{-\infty}^{\alpha_1} + \cdots + \int_{\alpha_m}^{\infty} f(R(x)) \, dx$$

$$= \pm \sum_{k=0}^{m} \int_{-\infty}^{\infty} f(s) x'_k(s) \, ds,$$

which is equal to

$$\int_{-\infty}^{\infty} f(x) \, dx.$$

$\square$

SOLUTION 6.9

The proof is essentially due to Gronwall (1918). Put

$$c_n = \int_{0}^{n\pi} \frac{1 - \cos x}{x} \, dx - \log(n\pi)$$

for any positive integer n. Then obviously

$$c_n = \int_{0}^{1} \frac{1 - \cos x}{x} \, dx - \int_{1}^{n\pi} \frac{\cos x}{x} \, dx,$$

and we are to show that c_n converges to Euler's constant γ as $n \to \infty$.

Problems and Solutions in Real Analysis

By the substitution $x = 2n\pi s$ we have

$$\int_0^{n\pi} \frac{1 - \cos x}{x} \, dx = \int_0^{1/2} \frac{1 - \cos 2n\pi s}{s} \, ds,$$

which is equal to

$$\pi \int_0^{1/2} \frac{1 - \cos 2n\pi s}{\sin \pi s} \, ds + \int_0^{1/2} \phi(s) \, ds - \int_0^{1/2} \phi(s) \cos 2n\pi s \, ds, \qquad (6.13)$$

where

$$\phi(s) = \frac{1}{s} - \frac{\pi}{\sin \pi s}$$

is a continuous function on the interval $[0, 1/2]$ if we define $\phi(0) = 0$. Hence by the remark after **PROBLEM 5.1** the third integral in (6.13) converges to 0 as $n \to \infty$. The second integral in (6.13) is equal to

$$\left[\log \frac{s}{\tan(\pi s / 2)} \right]_{s=0+}^{s=1/2} = \log \pi - 2 \log 2.$$

Finally, since it is easily verified that

$$\frac{1 - \cos 2n\pi s}{\sin \pi s} = 2 \sum_{k=1}^{n} \sin(2k - 1)\pi s,$$

the first integral in (6.13) is equal to

$$2\pi \sum_{k=1}^{n} \int_0^{1/2} \sin(2k - 1)\pi s \, ds = \sum_{k=1}^{n} \frac{2}{2k - 1},$$

which is $\log n + 2 \log 2 + \gamma + o(1)$ as $n \to \infty$. Thus we obtain $c_n = \gamma + o(1)$, which completes the proof. $\qquad \square$

Chapter 7

Series of Functions

- If each function $f_n(x)$ is continuous on a closed interval $[a, b]$ and if the series $\sum_{n=1}^{\infty} f_n(x)$ converges uniformly on $[a, b]$, then we have

$$\sum_{n=1}^{\infty} \int_a^b f_n(x)\, dx = \int_a^b \sum_{n=1}^{\infty} f_n(x)\, dx. \tag{7.1}$$

In other words, termwise integration is allowed.

- A simple and useful test for uniform convergence, known as Dirichlet's test, is as follows: Suppose that the Nth partial sum of the series $\sum f_n(x)$ is uniformly bounded (with respect to both N and x) on an interval I and that $g_n(x)$ is a monotone decreasing sequence converging uniformly to 0. Then the series

$$\sum_{n=1}^{\infty} f_n(x) g_n(x)$$

converges uniformly on I.

- If each function $f_n(x)$ is integrable over a closed interval $[a, b]$ in the sense of Riemann and if the nth partial sum of the series $\sum_{n=1}^{\infty} f_n(x)$ is uniformly bounded on $[a, b]$ and converges pointwise to the limit function which is also integrable over $[a, b]$, then (7.1) holds true. This is known as Arzelà's theorem.

- If each function $f_n(x)$ has the derivative $f_n'(x)$ at any point x in an open interval (a, b), if the series $\sum_{n=1}^{\infty} f_n(x)$ converges at least one point c in (a, b) and if $\sum_{n=1}^{\infty} f_n'(x)$ converges uniformly on (a, b) to a function $g(x)$, then $\sum_{n=1}^{\infty} f_n(x)$ converges uniformly on (a, b) and is differentiable at any point x in (a, b), whose derivative is equal to $g(x)$. Namely,

$$\left(\sum_{n=1}^{\infty} f_n(x) \right)' = \sum_{n=1}^{\infty} f_n'(x);$$

93

Problems and Solutions in Real Analysis

that is, termwise differentiation is allowed.

- An infinite series of the form

$$\sum_{n=0}^{\infty} a_n (z - z_0)^n$$

  with a complex z_0, a complex variable z and a complex sequence $\{a_n\}$, is called
  a *power series* about $z = z_0$.

- Given a power series, let

$$\frac{1}{\rho} = \limsup_{n\to\infty} |a_n|^{1/n}. \tag{7.2}$$

  Of course, we adopt the rule $\rho = 0$ or $\rho = \infty$ according as the limit superior on the right-hand side is equal to ∞ or 0 respectively. The number ρ is
  called the *radius of convergence* of the power series and (7.2) is referred to as
  Hadamard's formula.

  The circle $|z| = \rho$ called the circle of convergence has the following properties:

(a) The series converges absolutely and uniformly on compact sets in $|z| < \rho$.
(b) The sum is an analytic function and the derivative is obtained by termwise dif-
    ferentiation in $|z| < \rho$. The derived series has the same radius of convergence.
(c) If $|z| > \rho$, then the terms of the series are unbounded, and the series is divergent.

Note that nothing is claimed for the convergence on the circle. Tauber's theorem
(**PROBLEM 7.5**) describes the behavior on the circle of convergence.

- If $f(x)$ has derivatives of every order at any point a in an open interval I, the
  power series about $x = a$:

$$\sum_{n=0}^{\infty} \frac{f^{(n)}(a)}{n!} (x - a)^n$$

  is called the *Taylor series* generated by f.

  It then follows from Taylor's formula that this series represents the given function $f(x)$ if and only if the corresponding nth remainder term $R_n = R_n(x, a)$ converges to 0 as $n \to \infty$. For Lagrange's remainder term, see **SOLUTION 7.8**. Such
functions are called *real analytic functions*.

PROBLEM 7.1

Show that

$$\lim_{x \to 1-} \sqrt{1-x} \sum_{n=1}^{\infty} x^{n^2} = \frac{\sqrt{\pi}}{2}.$$

The behavior of the series

$$\sum_{n=1}^{\infty} z^{n^2}$$

when $z \to -1$ in a certain manner was used by Hardy (1914) to show that the Riemann zeta function $\zeta(z)$ has an infinitely many zeros on the critical line. This proof was materially simplified by Landau (1915), who showed that no property of this series was needed for the purpose of the proof except for the upper estimate

$$O\left((1-|z|)^{-1/2}\right).$$

PROBLEM 7.2

Show that

$$\lim_{x \to 1-} (1-x)^2 \sum_{n=1}^{\infty} \frac{n x^n}{1-x^n} = \frac{\pi^2}{6}.$$

The series of the form

$$\sum_{n=1}^{\infty} \frac{a_n x^n}{1-x^n}$$

is called the Lambert series and transformed (formally) to

$$\sum_{n=1}^{\infty} \left(\sum_{d \mid n} a_d\right) x^n,$$

where d runs over all divisors of n. The reader may be enticed to find any other example of a simple power series $f(x)$ for which the limit of $(1-x)^\alpha f(x)$, as $x \to 1-$, is equal to a rational multiple of π^α.

PROBLEM 7.3

Suppose that a power series

$$f(x) = \sum_{n=0}^{\infty} a_n x^n$$

has the radius of convergence $\rho > 0$ and that $\sum_{n=0}^{\infty} a_n \rho^n$ converges to α. Show that $f(x)$ converges to α as $x \to \rho-$.

This is known as Abel's continuity theorem (1826), which was published in the first volume of Crelle Journal. This is the first major mathematical journal, except for proceedings of academies, founded by August Leopold Crelle.

PROBLEM 7.4

Suppose that a power series

$$g(x) = \sum_{n=0}^{\infty} b_n x^n$$

has the radius of convergence $\rho > 0$, all b_n are positive, and that $\sum_{n=0}^{\infty} b_n \rho^n$ diverges to ∞. Then show that $f(x)/g(x)$ converges to α as $x \to \rho-$ for any power series

$$f(x) = \sum_{n=0}^{\infty} a_n x^n$$

such that a_n/b_n converges to α as $n \to \infty$.

This is due to Cesàro (1893), and is a generalization of Abel's continuity theorem. For, applying Cesàro's theorem to $a_n = c_0 + c_1 + \cdots + c_n$ and $b_n = 1$ for a convergent series $\sum_{n=0}^{\infty} c_n$, we get

$$\sum_{n=0}^{\infty} c_n x^n = \frac{a_0 + a_1 x + \cdots + a_n x^n + \cdots}{1 + x + \cdots + x^n + \cdots} \to \sum_{n=0}^{\infty} c_n$$

as $x \to 1-$.

Cesàro's theorem can be used to compute the limit in **PROBLEM 7.1** as follows: Put

$$f(x) = \sum_{n=1}^{\infty} x^{n^2} = (1-x) \sum_{n=1}^{\infty} [\sqrt{n}] x^n = (1-x) F(x)$$

and

$$G(x) = (1-x)^{-3/2} = 1 + \sum_{n=1}^{\infty} b_n x^n$$

where

$$b_n = \frac{1}{n!} \cdot \frac{3}{2} \cdots \left(n + \frac{1}{2}\right) \sim 2\sqrt{n/\pi}$$

as $n \to \infty$. This asymptotic formula follows from **PROBLEM 16.1** and $\Gamma(1/2) = \sqrt{\pi}$ where $\Gamma(s)$ is the Gamma function. We then have

$$\lim_{x \to 1-} \sqrt{1-x}\, f(x) = \lim_{x \to 1-} \frac{F(x)}{G(x)} = \frac{\sqrt{\pi}}{2}$$

since $[\sqrt{n}]/b_n$ converges to $\sqrt{\pi}/2$ as $n \to \infty$.

PROBLEM 7.5

Suppose that a power series $f(x) = \sum_{n=0}^{\infty} a_n x^n$ has the radius of convergence $\rho > 0$, na_n converges to 0 as $n \to \infty$, and that $f(x)$ converges to α as $x \to \rho-$. Show then that

$$\sum_{n=0}^{\infty} a_n \rho^n = \alpha.$$

This is known as Tauber's theorem (1897). Partial converses to Abel's continuity theorem are generally called Tauberian theorems. See the comment after SOLUTION 7.5.

PROBLEM 7.6

Suppose that a power series

$$f(x) = \sum_{n=0}^{\infty} a_n x^n$$

has the radius of convergence 1, all a_n are non-negative, and that $(1-x)f(x)$ converges to 1 as $x \to 1-$. Show then that

$$\lim_{n \to \infty} \frac{a_0 + a_1 + a_2 + \cdots + a_n}{n} = 1.$$

Karamata (1930) gave an elegant proof using Weierstrass' approximation theorem, which was a new proof to Littlewood's rather difficult theorem (1910) stated in SOLUTION 7.5. According to Nikolić (2002) Karamata's two-page paper created a sensation in mathematical circles. See also Wielandt (1952).

PROBLEM 7.7

Show that the series

$$f(x) = \sum_{n=0}^{\infty} e^{-n} \cos n^2 x$$

is infinitely differentiable everywhere, but the Taylor series about $x = 0$

$$\sum_{n=0}^{\infty} \frac{f^{(n)}(0)}{n!} x^n$$

does not converge except for the origin.

PROBLEM 7.8

Suppose that $f \in C^{\infty}(\mathbb{R})$ satisfies $f^{(n)}(x) \geq 0$ for any non-negative integer n and for all $x \in \mathbb{R}$. Show that the Taylor series about $x = 0$ generated by f converges for all x.

For example, the function $f(x) = a^x$ for $a > 1$ satisfies the condition stated in the problem.

This is a special case of the result obtained by Bernstein (1928). Bernstein's theorem on a finite interval is explained in the book of Apostol (1957) on p. 418.

PROBLEM 7.9

Let $\{a_n\}$ be a monotone decreasing sequence converging to 0. Show that the trigonometric series

$$\sum_{n=1}^{\infty} a_n \sin n\theta$$

converges uniformly on $\mathbb{R}$ if and only if na_n converges to 0 as $n \to \infty$.

This is due to Zygmund (1979) on p. 182. In particular, if a trigonometric series represents a discontinuous functions, then na_n does not converge to 0. This is illustrated by the following example.

PROBLEM 7.10

Show that the trigonometric series

$$\sum_{n=1}^{\infty} \frac{\sin n\theta}{n}$$

converges uniformly to

$$\frac{\pi - \theta}{2}$$

on the interval $[\delta, 2\pi - \delta]$ for any $\delta > 0$.

This is the Fourier expansion for the first periodic Bernoulli polynomial defined by $\overline{B}_1(x) = x - [x] - 1/2$ $(x \notin \mathbb{Z})$. See also **PROBLEM 1.7** and **PROBLEM 5.9**.

Series of Functions   99

PROBLEM 7.11

Suppose that $f \in C^1(0, 1)$ and $\displaystyle\int_0^1 |f(x)|\,dx$ converges. Show that the Fourier series

$$\frac{a_0}{2} + \sum_{n=1}^{\infty} (a_n \cos 2n\pi x + b_n \sin 2n\pi x)$$

converges to $f(x)$ in the interval $(0, 1)$.

The Fourier coefficients of $f(x)$ on the interval $(0, 1)$ are defined by

$$a_n = 2 \int_0^1 f(x) \cos 2n\pi x\,dx \quad \text{and} \quad b_n = 2 \int_0^1 f(x) \sin 2n\pi x\,dx.$$

PROBLEM 7.12

Let

$$\sum_{n=1}^{\infty} a_n x^n$$

be the Taylor series about $x = 0$ of the algebraic function

$$f(x) = \frac{1 - \sqrt{1 - 4x}}{2}.$$

Show that each a_n is a positive integer and that a_n is odd if and only if n is a power of 2.

The first sixteen coefficients are as follows:

$$\underline{1}, \quad \underline{1}, \quad 2, \quad \underline{5}, \quad 14, \quad 42, \quad 132, \quad \underline{429}, \quad 1430, \quad 4862,$$

$$16796, \quad 58786, \quad 208012, \quad 742900, \quad 2674440, \quad \underline{9694845}$$

where odd coefficients are indicated by underlines.

Solutions for Chapter 7

SOLUTION 7.1

Applying **PROBLEM 6.4** for $f(x) = e^{-x^2}$ we have

$$h \sum_{n=0}^{\infty} \exp\left(-n^2 h^2\right) \to \int_0^{\infty} e^{-x^2} \, dx = \frac{\sqrt{\pi}}{2}$$

as $h \to 0+$. By the substitution $h = \sqrt{-\log x}$ it follows that h converges to $0+$ if and only if x converges to $1-$; hence, using $-\log x \sim 1 - x$ as $x \to 1-$, we obtain

$$\lim_{x \to 1-} \sqrt{1-x} \sum_{n=0}^{\infty} x^{n^2} = \lim_{x \to 1-} \sqrt{-\log x} \sum_{n=0}^{\infty} x^{n^2}$$

$$= \lim_{h \to 0+} h \sum_{n=0}^{\infty} \exp\left(-n^2 h^2\right) = \frac{\sqrt{\pi}}{2}.$$

$\square$

SOLUTION 7.2

Applying **PROBLEM 6.4** for $f(x) = x/(e^x - 1)$ we get

$$h \sum_{n=1}^{\infty} \frac{nh}{e^{nh} - 1} \to \int_0^{\infty} \frac{x \, dx}{e^x - 1} = \frac{\pi^2}{6}$$

as $h \to 0+$. By the substitution $h = -\log x$ it holds that h converges to $0+$ if and only if x converges to $1-$; hence, using $-\log x \sim 1 - x$ as $x \to 1-$,

$$\lim_{x \to 1-} (1-x)^2 \sum_{n=1}^{\infty} \frac{nx^n}{1 - x^n} = \lim_{x \to 1-} (\log x)^2 \sum_{n=1}^{\infty} \frac{nx^n}{1 - x^n}$$

$$= \lim_{h \to 0+} h \sum_{n=1}^{\infty} \frac{nh}{e^{nh} - 1} = \frac{\pi^2}{6}.$$

Note that the function $f(x) = x/(e^x - 1)$ can be regarded as a continuous function on the interval $[0, \infty)$ if we define $f(0) = 1$.

$\square$

SOLUTION 7.3

Replacing x by ρx we can assume that $\rho = 1$ without loss of generality. Put

$$s_n = a_0 + a_1 + \cdots + a_n.$$

Since s_n converges to α as $n \to \infty$, for any $\epsilon > 0$ we can take a sufficiently large integer N satisfying

$$|s_n - \alpha| < \epsilon$$

for all integers $n > N$. For $0 < x < 1$ we have

$$\frac{f(x)}{1-x} = \sum_{n=0}^{\infty} s_n x^n = \frac{\alpha}{1-x} + \sum_{n=0}^{\infty}(s_n - \alpha)x^n \,;$$

therefore

$$|f(x) - \alpha| \le (1-x)\sum_{n=0}^{N}|s_n - \alpha|x^n + (1-x)\sum_{n>N}|s_n - \alpha|x^n$$

$$< (1-x)\sum_{n=0}^{N}|s_n - \alpha| + (1-x)\sum_{n=0}^{\infty}\epsilon x^n$$

$$= (1-x)\sum_{n=0}^{N}|s_n - \alpha| + \epsilon.$$

The right-hand side can be $< 2\epsilon$ by letting x be sufficiently close to $1-$. This means that $f(x)$ converges to α as $x \to 1-$. $\qquad\square$

SOLUTION 7.4

As in the previous problem we can assume that $\rho = 1$. For any $\epsilon > 0$ there is a positive integer N such that $|a_n - \alpha b_n| < \epsilon b_n$ for all n greater than N. Since

$$f(x) = \sum_{n=0}^{\infty} a_n x^n = \alpha g(x) + \sum_{n=0}^{\infty}(a_n - \alpha b_n)x^n,$$

we have

$$\left|\frac{f(x)}{g(x)} - \alpha\right| \le \frac{1}{g(x)}\left(\sum_{n=0}^{N}|a_n - \alpha b_n| + \epsilon \sum_{n>N} b_n x^n\right)$$

$$< \frac{1}{g(x)}\sum_{n=0}^{N}|a_n - \alpha b_n| + \epsilon \qquad\qquad (7.3)$$

for any $0 < x < 1$. Now $g(x)$ diverges to ∞ as $x \to 1-$, since

$$\liminf_{x \to 1-} g(x) \geq \sum_{k=0}^{n} b_k$$

for any positive integer n. Hence the right-hand side of (7.3) can be smaller than 2ϵ if we take x sufficiently close to 1. □

SOLUTION 7.5

As in the previous problems we can assume that $\rho = 1$. Put

$$s_n = a_0 + a_1 + \cdots + a_n.$$

For any $\epsilon > 0$ there is a positive integer N such that $n|a_n| < \epsilon$ for all n greater than N. For any $0 < x < 1$ and any $n > N$ we obtain

$$|s_n - f(x)| \leq \sum_{k=1}^{n} |a_k|(1 - x^k) + \sum_{k>n} k|a_k| \frac{x^k}{k}$$

$$\leq (1 - x) \sum_{k=1}^{n} k|a_k| + \frac{\epsilon}{n(1 - x)}.$$

Substituting $x = 1 - 1/n$ we infer that

$$\left| s_n - f\left(1 - \frac{1}{n}\right) \right| \leq \frac{|a_1| + 2|a_2| + \cdots + n|a_n|}{n} + \epsilon.$$

Therefore the right-hand side can be smaller than 2ϵ if we take n sufficiently large. This means that s_n converges to α as $n \to \infty$. □

REMARK. Pringsheim (1900) weakened Tauber's condition $na_n = o(1)$ to

$$a_1 + 2a_2 + \cdots + na_n = o(n)$$

as $n \to \infty$. To see this we put $\tau_0 = 0$ and $\tau_n = a_1 + 2a_2 + \cdots + na_n$. Since $a_n = (\tau_n - \tau_{n-1})/n$ converges to 0 as $n \to \infty$, the given series $f(x)$ converges in

$|x| < 1$. Moreover for any $0 < x < 1$ we obtain

$$f(x) = a_0 + \sum_{n=1}^{\infty} \frac{\tau_n - \tau_{n-1}}{n} x^n$$

$$= a_0 + \sum_{n=1}^{\infty} \tau_n \left(\frac{x^n}{n} - \frac{x^{n+1}}{n+1} \right)$$

$$= a_0 + (1 - x) \sum_{n=1}^{\infty} \frac{\tau_n}{n+1} x^n + \sum_{n=1}^{\infty} \frac{\tau_n}{n(n+1)} x^n.$$

Since $\tau_n = o(n)$, the second term on the right-hand side clearly converges to 0 as $x \to 1-$. This implies that the third term converges to $\alpha - a_0$ as $x \to 1-$. Since $\tau_n/(n(n+1)) = o(1/n)$ as $n \to \infty$, we can apply Tauber's theorem to this power series so that

$$\alpha - a_0 = \lim_{n \to \infty} \sum_{k=1}^{n} \frac{\tau_k}{k(k+1)}$$

$$= \lim_{n \to \infty} \left(\tau_1 + \frac{\tau_2 - \tau_1}{2} + \cdots + \frac{\tau_n - \tau_{n-1}}{n} - \frac{\tau_n}{n+1} \right)$$

$$= \sum_{n=1}^{\infty} a_n,$$

as required.

Furthermore Littlewood (1910) has shown that Tauber's theorem holds true if the sequence na_n is bounded. Finally Hardy and Littlewood (1914) proved it even if na_n is either bounded above or bounded below. To see this we need the following result.

SOLUTION 7.6

It follows from the assumption that

$$(1 - x) \sum_{n=0}^{\infty} a_n x^{(k+1)n} = \frac{1}{1 + x + \cdots + x^k} (1 - x^{k+1}) \sum_{n=0}^{\infty} a_n \left(x^{k+1} \right)^n$$

converges to

$$\frac{1}{k+1} = \int_0^1 t^k \, dt$$

as $x \to 1-$ for any non-negative integer k. Therefore we have

$$\lim_{x \to 1-} (1 - x) \sum_{n=0}^{\infty} a_n x^n P(x^n) = \int_0^1 P(t) \, dt$$

for any polynomial $P(t)$.

We now introduce the discontinuous function $\phi(x)$ defined by

$$\phi(x) = \begin{cases} 0 & \text{for } 0 \le x < 1/e, \\ 1/x & \text{for } 1/e \le x \le 1. \end{cases}$$

For any $\epsilon > 0$ we can find two continuous functions $\phi_\pm(x)$ defined on the interval $[0, 1]$ such that $\phi_-(x) \le \phi(x) \le \phi_+(x)$ and $\phi_+(x) - \phi_-(x) < \epsilon$ for any x in $[0, 1]$. By Weierstrass' approximation theorem there are polynomials $P_\pm(x)$ satisfying $|\phi_\pm(x) \pm \epsilon - P_\pm(x)| < \epsilon$ respectively. Hence it follows that $P_-(x) < \phi(x) < P_+(x)$ and $P_+(x) - P_-(x) < 5\epsilon$.

Since a_n are all non-negative, we obtain

$$\sum_{n=0}^{\infty} a_n x^n P_-(x^n) \le \sum_{n=0}^{\infty} a_n x^n \phi(x^n) \le \sum_{n=0}^{\infty} a_n x^n P_+(x^n)$$

for any $0 < x < 1$. If we put $x = e^{-1/N}$, then $x \to 1-$ if and only if $N \to \infty$; hence, using $1 - e^{-1/N} \sim 1/N$, we have

$$\int_0^1 P_-(t)\,dt \le \liminf_{N\to\infty} \frac{1}{N} \sum_{n=0}^{N} a_n$$

and

$$\limsup_{N\to\infty} \frac{1}{N} \sum_{n=0}^{N} a_n \le \int_0^1 P_+(t)\,dt.$$

Therefore, since $P_+(x) - P_-(x) < 5\epsilon$ and

$$\int_0^1 \phi(x)\,dx = \int_{1/e}^1 \frac{dx}{x} = 1,$$

it follows that

$$1 < \int_0^1 P_+(t)\,dt < \int_0^1 P_-(t)\,dt + 5\epsilon < 1 + 5\epsilon.$$

Hence the sequence

$$\frac{1}{N} \sum_{n=0}^{N} a_n$$

converges to 1 as $N \to \infty$. $\qquad\qquad\qquad\qquad\qquad\qquad\qquad\qquad\square$

Remark. Using this result we can show that $\tau_n = o(n)$ as $n \to \infty$ even if $na_n < K$ for some positive constant K, with τ_n as defined in the remark after **Solution 7.5**. This gives a simpler proof of the theorem due to Hardy and Littlewood (1914).

First of all we have

$$f''(x) = \sum_{n=2}^{\infty} n(n-1)a_n x^{n-2}$$

$$\leq K \sum_{n=2}^{\infty} (n-1)x^{n-2} = \frac{K}{(1-x)^2}$$

for any $0 < x < 1$. Put

$$g(t) = f(1 - e^{-t}) \in C^{\infty}(0, \infty)$$

for brevity. By assumption $g(t)$ converges to α as $t \to \infty$. Moreover

$$g''(t) + g'(t) = e^{-2t}f''(1 - e^{-t}) \leq K$$

holds for any positive t. It follows from **Problem 4.9** that $g'(t)$ converges to 0. This means that $(1-x)f'(x)$ converges to 0 as $x \to 1-$. Then the power series

$$\sum_{n=1}^{\infty}\left(1 - \frac{na_n}{K}\right)x^{n-1} = \frac{1}{1-x} - \frac{f'(x)}{K}$$

satisfies all the conditions stated in **Problem 7.6**; therefore

$$\frac{1}{N}\sum_{n=1}^{N}\left(1 - \frac{na_n}{K}\right) = 1 - \frac{\tau_N}{KN}$$

converges to 1 as $N \to \infty$. Hence $\tau_n = o(n)$ as $n \to \infty$.

$\boxed{\text{Solution 7.7}}$

By k times termwise differentiation of the given series we get

$$\sum_{n=1}^{\infty} n^{2k}e^{-n}\,\mathfrak{R}\left(i^k e^{in^2 x}\right),$$

which clearly converges uniformly on $\mathbb{R}$. Hence this series represents $f^{(k)}(x)$; in particular,

$$f^{(k)}(0) = \mathfrak{R}(i^k)\sum_{n=1}^{\infty} n^{2k}e^{-n}.$$

We thus have

$$|f^{(2\ell)}(0)| \geq \left(\frac{4\ell}{e}\right)^{4\ell}$$

for any positive integer ℓ by looking at only the 4ℓth term. Let ρ be the radius of convergence of $f(x)$. It then follows from Hadamard's formula (7.2) that

$$\frac{1}{\rho} \geq \limsup_{\ell \to \infty} \left(\frac{(4\ell)^{4\ell}}{(2\ell)!\, e^{4\ell}}\right)^{1/(2\ell)} \geq \limsup_{\ell \to \infty} \frac{(4\ell)^2}{2e^2\ell} = \infty,$$

which means $\rho = 0$. □

SOLUTION 7.8

It follows from Taylor's formula about $x = a$ with Lagrange's remainder term that

$$f(x) = \sum_{k=0}^{n} \frac{f^{(k)}(a)}{k!}(x-a)^k + \frac{f^{(n+1)}(\xi)}{(n+1)!}(x-a)^{n+1}$$

for some ξ between x and a. If we take $x = 2a > 0$, then

$$f(2a) = \sum_{k=0}^{n} \frac{f^{(k)}(a)}{k!} a^k + \frac{f^{(n+1)}(\xi)}{(n+1)!} a^{n+1}$$

$$\geq \sum_{k=0}^{n} \frac{f^{(k)}(a)}{k!} a^k,$$

which implies the convergence of the series

$$\sum_{n=0}^{\infty} \frac{f^{(n)}(a)}{n!} a^n.$$

In particular, $f^{(n)}(a)a^n/n!$ converges to 0 as $n \to \infty$ for any $a > 0$. Hence

$$\left| f(x) - \sum_{k=0}^{n} \frac{f^{(k)}(0)}{k!} x^k \right| = \frac{f^{(n+1)}(\xi)}{(n+1)!} |x|^{n+1}$$

$$\leq \frac{f^{(n+1)}(|x|)}{(n+1)!} |x|^{n+1} \to 0$$

as $n \to \infty$, since by assumption the derivative of $f(x)$ of any order is monotone increasing for any x. □

SOLUTION 7.9

Suppose first that the given series converges uniformly on $\mathbb{R}$. For any $\epsilon > 0$ we can take a positive integer N such that

$$\max_{\theta \in \mathbb{R}} \left| a_p \sin p\theta + a_{p+1} \sin (p+1)\theta + \cdots + a_q \sin q\theta \right| < \epsilon$$

for any integers $q > p$ greater than N. We now choose $\theta = \pi/(4p)$ and $q = 2p$ so that $a_k \sin k\theta \geq 0$ for $p \leq k \leq 2p$. Then we have

$$\epsilon > \sum_{n=p}^{2p} a_n \sin \frac{n\pi}{4p} \geq \frac{1}{\sqrt{2}} \sum_{n=p}^{2p} a_n \geq \frac{p+1}{\sqrt{2}} a_{2p}.$$

Hence

$$\max\{2pa_{2p}, (2p+1)a_{2p+1}\} \leq 2(p+1)a_{2p} < 2\sqrt{2}\,\epsilon,$$

and so $na_n \to 0$ as $n \to \infty$.

Conversely assume that na_n converges to 0 as $n \to \infty$. For any $\epsilon > 0$ we can take a positive integer N so that $na_n < \epsilon$ for any integer n greater than N. For any $\theta \in (0, \pi]$ let m_θ be a unique positive integer satisfying $\pi/(m+1) < \theta \leq \pi/m$. For any integers $q > p$ greater than N let $S_0(\theta)$ and $S_1(\theta)$ be the sums of $a_n \sin n\theta$ from $n = p$ to r and from $n = r+1$ to q respectively, where

$$r = \min\{q, p + m_\theta - 1\}.$$

For the sum $S_0(\theta)$ we use an almost trivial estimate $\sin x < x$ for $x > 0$ to obtain

$$|S_0(\theta)| \leq \theta \sum_{n=p}^{r} na_n < \theta m_\theta \epsilon \leq \pi\epsilon.$$

Next for the sum $S_1(\theta)$ we can assume $q \geq p + m_\theta$; so $r = p + m_\theta - 1$. By partial summation we have

$$|S_1(\theta)| = \left| \sum_{n=r+1}^{q} a_n(\sigma_n - \sigma_{n-1}) \right|$$

$$\leq a_{r+1}|\sigma_r| + a_q|\sigma_q| + (a_{r+1} - a_q) \max_{r<k<q} |\sigma_k|$$

where

$$\sigma_n = \sin\theta + \sin 2\theta + \cdots + \sin n\theta.$$

Using Jordan's inequality $\pi \sin x \geq 2x$ valid on the interval $[0, \pi/2]$, we get

$$|\sigma_n| = \left| \frac{\cos \vartheta - \cos(2n+1)\vartheta}{\sin \vartheta} \right| \leq \frac{2}{\sin \vartheta} \leq \frac{\pi}{\theta} < m_\theta + 1$$

where $\theta = 2\vartheta$; therefore

$$|S_1(\theta)| \leq (m_\theta + 1)(a_{r+1} + a_q + a_{r+1} - a_q)$$
$$\leq 2(r+1)a_{r+1} < 2\epsilon.$$

We then have

$$\left| \sum_{n=p}^{q} a_n \sin n\theta \right| \leq |S_0(\theta)| + |S_1(\theta)| < (\pi + 2)\epsilon,$$

which holds uniformly on the interval $[0, \pi]$; whence on $\mathbb{R}$ by symmetry and periodicity. □

SOLUTION 7.10

The uniform convergence on any interval $[\delta, 2\pi - \delta]$ with $\delta > 0$ can be easily derived from Dirichlet's test (See Item 2 on p. 93). Thus it suffices to show that the limit function is $(\pi - \theta)/2$.

Put $\theta = 2\vartheta$ for brevity. Integrating the formula

$$\frac{1}{2} + \sum_{n=1}^{m} \cos n\theta = \frac{\sin(2m+1)\vartheta}{2 \sin \vartheta}$$

from 0 to $\omega \in [\delta, 2\pi - \delta]$, we obtain

$$\frac{\omega}{2} + \sum_{n=1}^{m} \frac{\sin n\omega}{n} = \int_0^\omega \frac{\sin(2m+1)\vartheta}{2 \sin \vartheta} d\theta$$
$$= \int_0^\eta \frac{\sin(2m+1)\vartheta}{\sin \vartheta} d\vartheta$$

where $\omega = 2\eta$. We divide the last integral into two parts,

$$\int_0^\eta \frac{\sin(2m+1)t}{t} dt + \int_0^\eta \phi(t) \sin(2m+1)t \, dt$$

and call them $I_m(\eta)$ and $J_m(\eta)$ respectively, where

$$\phi(t) = \frac{1}{\sin t} - \frac{1}{t}.$$

Note that $\phi \in C^1(-\pi, \pi)$ if we define $\phi(0) = 0$.

Integrating by parts we get

$$J_m(\eta) = -\frac{\phi(\eta)}{2m+1} \cos(2m+1)\eta$$

$$+ \frac{1}{2m+1} \int_0^\eta \phi'(t) \cos(2m+1)t \, dt,$$

from which it is easily seen that

$$|J_m(\eta)| < \frac{M_0 + \pi M_1}{2m+1}$$

where M_0 and M_1 are the maxima of $|\phi|$ and $|\phi'|$ on the interval $[0, \pi - \delta/2]$ respectively. Thus $J_m(\eta)$ converges uniformly to 0 as $m \to \infty$ on $[0, \pi - \delta/2]$.

We next deal with the integral $I_m(\eta)$. By the substitution $s = (2m+1)t$ we have

$$I_m(\eta) = \int_0^{(2m+1)\eta} \frac{\sin s}{s} \, ds.$$

Since $\pi/2 = I_m(\pi/2) + J_m(\pi/2)$, we have for any $x > \pi$

$$\left| \frac{\pi}{2} - \int_0^x \frac{\sin s}{s} \, ds \right| \leq \left| J_{m_x}\left(\frac{\pi}{2}\right) \right| + \int_{(2m_x+1)\pi/2}^x \frac{ds}{s}$$

$$\leq \left| J_{m_x}\left(\frac{\pi}{2}\right) \right| + \frac{1}{2m_x+1},$$

where m_x is the largest integer satisfying $(2m+1)\pi \leq 2x$. Since the right-hand side converges to 0 as $x \to \infty$, the improper integral

$$\int_0^\infty \frac{\sin s}{s} \, ds$$

exists and equals to $\pi/2$. Therefore

$$\left| \frac{\omega - \pi}{2} + \sum_{n=1}^m \frac{\sin n\omega}{n} \right| \leq |J_m(\eta)| + \left| I_m(\eta) - \frac{\pi}{2} \right|$$

$$= |J_m(\eta)| + \left| \int_{(2m+1)\eta}^\infty \frac{\sin s}{s} \, ds \right|.$$

The right-hand side clearly converges to 0 uniformly in $\eta \in [\delta/2, \pi - \delta/2]$ as $m \to \infty$. $\qquad \square$

SOLUTION 7.11

For an arbitrary fixed $x \in (0, 1)$ let $s_n(x)$ be nth partial sum of the Fourier series. Then we have

$$
s_n(x) = \frac{a_0}{2} + \sum_{k=1}^{n} (a_k \cos 2k\pi x + b_k \sin 2k\pi x)
$$

$$
= \int_0^1 f(t) \left(1 + 2 \sum_{k=1}^{n} \cos 2k\pi(t - x)\right) dt.
$$

Now, as we have already seen in **SOLUTION 7.10**, the trigonometric sum in the big parentheses can be expressed in the ratio of sines; hence we obtain

$$
s_n(x) = \int_{-x}^{1-x} f(x + y) \frac{\sin(2n + 1)\pi y}{\sin \pi y} \, dy.
$$

If $f(x) = 1$, then obviously $a_0 = 1$ and $a_n = b_n = 0$ for all $n \geq 1$; hence

$$
1 = \int_{-x}^{1-x} \frac{\sin(2n + 1)\pi y}{\sin \pi y} \, dy.
$$

Therefore we have

$$
s_n(x) - f(x) = \int_{-x}^{1-x} \phi_x(y) \sin(2n + 1)\pi y \, dy
$$

where

$$
\phi_x(y) = \frac{f(x + y) - f(x)}{\sin \pi y}.
$$

Clearly $\phi_x \in C(-x, 1 - x)$ and the improper integral

$$
\int_{-x}^{1-x} |\phi_x(y)| \, dy
$$

converges, since $f \in C^1(0, 1)$ and $\int_0^1 |f(x)| \, dx$ converges. Thus it follows from the remark after **SOLUTION 5.1** that

$$
\lim_{n \to \infty} s_n(x) = f(x),
$$

in view of

$$
\int_x^{1-x} \sin 2\pi y \, dy = \int_x^{1-x} \cos 2\pi y \, dy = 0.
$$

□

SOLUTION 7.12

It follows from

$$\left(1 - 2\sum_{n=1}^{\infty} a_n x^n\right)^2 = 1 - 4x$$

that $a_1 = 1$ and the recursion formula

$$a_{n+1} = a_1 a_n + a_2 a_{n-1} + \cdots + a_n a_1$$

holds for any positive integer n. This implies immediately that every a_n is a positive integer; therefore

$$a_{2k+1} = 2(a_1 a_{2k} + \cdots + a_k a_{k+1})$$

is an even integer for any positive integer k.

Suppose now that there exists a non-negative integer ℓ such that a_n is even for every $n = 2^\ell(2k + 1)$ with $k \geq 1$. Then

$$a_{2n} = 2(a_1 a_{2n-1} + a_2 a_{2n-2} + \cdots + a_{n-1} a_{n+1}) + a_n^2$$

implies that a_{2n} is also even for every

$$n = 2^\ell(2k + 1)$$

with $k \geq 1$. Hence by induction every a_n is shown to be even except for the case in which n is a power of 2. However, if n is a power of 2, we can similarly show that a_n is odd, since a_1 is odd.   □

REMARK. By the Taylor expansion of $\sqrt{1 + x}$ it is easily seen that

$$a_n = \frac{1}{n}\binom{2n - 2}{n - 1}.$$

We adopt, of course, the convention that $0! = 1$. Thus the above result may give some information on the divisibility of the central binomial coefficients.

.

Chapter 8

Approximation by Polynomials

- Any continuous function $f(x)$ on a bounded closed interval $[a, b]$ can be approximated uniformly by polynomials; that is, for any $\epsilon > 0$ there exists a polynomial $P(x)$ satisfying

$$|f(x) - P(x)| < \epsilon$$

for all x in $[a, b]$. This is known as Weierstrass' approximation theorem (1885) and may be easily shown in an elementary way by using the Bernstein polynomials (See **PROBLEM 8.1**).

- Any continuous periodic function $f(x)$ with period 2π can be approximated uniformly by trigonometric polynomials; that is, for any $\epsilon > 0$ there exists a trigonometric polynomial

$$P(\theta) = \sum_{n=0}^{m} (a_n \cos n\theta + b_n \sin n\theta)$$

satisfying $|f(\theta) - P(\theta)| < \epsilon$ for all θ (See also **PROBLEM 8.3**). This theorem is referred to as the approximation theorem by trigonometric polynomials.

The latter may be derived from Fejér's summability theorem, a very important result in the theory of Fourier series, which shows us a constructive way for trigonometric polynomials:

$$P(\theta) = \frac{1}{m}(s_0(\theta) + s_1(\theta) + \cdots + s_{m-1}(\theta))$$

where $s_n(\theta)$ is the nth partial sum of the Fourier series for $f(x)$.

PROBLEM 8.1

For any $f \in C[0, 1]$ the polynomial of degree n defined by

$$B_n(f;x) = \sum_{k=0}^{n} f\left(\frac{k}{n}\right)\binom{n}{k} x^k(1-x)^{n-k}$$

is called the **Bernstein polynomial**. Show then that $B_n(f;x)$ converges to $f(x)$ uniformly on the interval $[0, 1]$.

Runge (1885ab) gave nearly simultaneous proof for Weierstrass' approximation theorem (1885), but his papers do not explicitly contain Weierstrass' theorem. This fact was pointed out by L. E. Phragmén in the paper of Mittag-Leffler (1900).

Many other proofs of the approximation theorem appeared shortly after Weierstrass. Picard (1891) used the Poisson integral and Volterra (1897) used Dirichlet's principle. Lebesgue (1898) used essentially the following uniformly convergent series (not Taylor's series in x) on the interval $[-1, 1]$:

$$|x| = 1 - \frac{1}{2}(1-x^2) - \frac{1}{2\cdot4}(1-x^2)^2 - \frac{1\cdot3}{2\cdot4\cdot6}(1-x^2)^3 - \cdots.$$

This is Lebesgue's first paper. In a letter to É. Picard, Mittag-Leffler (1900) gave an elementary proof using the following discontinuity property:

$$\lim_{n\to\infty} \chi_n(x) = \begin{cases} 1 & \text{for} & x > 0, \\ 0 & \text{for} & x = 0, \\ -1 & \text{for} & -2 < x < 0, \end{cases}$$

where $\chi_n(x) = 1 - 2^{1-(1+x)^n}$. Fejér (1900) showed that the Fourier series for a bounded integrable function is uniformly Cesàro summable of the first order on an interval on which the function is continuous. (In this paper his name was misprinted to 'Tejér'.) Fejér was 20 years old when he wrote this paper and obtained the doctoral thesis at Univ. of Budapest under H. A. Schwarz 2 years later. Lerch (1903) gave a proof using the Fourier series of special but simple piecewise linear functions. Landau (1908) used the formula:

$$\lim_{n\to\infty} \frac{\displaystyle\int_0^1 f(x)(1-(x-y)^2)^n\,dx}{\displaystyle\int_0^1 (1-x^2)^n\,dx} = f(y).$$

Bernstein (1912a) gave a probabilistic proof of Weierstrass' approximation theorem by introducing the Bernstein polynomials. Carleman (1927) showed that, for any $f \in C(\mathbb{R})$, there exists a sequence of entire functions converging to $f(x)$ uniformly on $\mathbb{R}$.

PROBLEM 8.2

Show that there exists a sequence of polynomials with integral coefficients converging to $f \in C[0, 1]$ uniformly if $f(0) = f(1) = 0$.

Pál (1914) noted that for any $\epsilon > 0$ there exists a sequence of polynomials with integral coefficients converging uniformly to $f \in C[\epsilon - 1, 1 - \epsilon]$ if $f(0) = 0$. Kakeya (1914) found necessary and sufficient conditions on $f \in C[-1, 1]$ which can be approximated by polynomials with integer coefficients. Pál's work was extended to larger intervals by Kakeya (1914) and Okada (1923).

PROBLEM 8.3

Deduce the approximation theorem by trigonometric polynomials from Weierstrass' approximation theorem by polynomials.

PROBLEM 8.4

For any $f \in C^1[0, 1]$ show that $B'_n(f; x)$ converges to $f'(x)$ uniformly on the interval $[0, 1]$, where $B_n(f; x)$ is the Bernstein polynomial.

PROBLEM 8.5

Show that $f \in C[0, 1]$ satisfies

$$\int_0^1 x^n f(x) \, dx = 0$$

for all non-negative integer n if and only if $f(x)$ vanishes everywhere on the interval $[0, 1]$.

We can replace the condition $f \in C[0, 1]$ by a weaker one $f \in C(0, 1)$ supposing in addition that the improper integral

$$\int_0^1 f(x) \, dx$$

converges absolutely. To see this consider

$$F(x) = \int_0^x f(t) \, dt.$$

Plainly $F \in C[0, 1]$, $F(0) = 0$ and

$$\int_0^1 x^n F(x) \, dx = \left[\frac{x^{n+1} - 1}{n + 1} F(x) \right]_{x=0}^{x=1} - \frac{1}{n + 1} \int_0^1 (x^{n+1} - 1) f(x) \, dx = 0$$

for any non-negative integer n, which implies the vanishing of f.

It is easily seen that the interval $[0, 1]$ can be replaced by any compact interval using a suitable affine transformation. However we cannot replace the interval $[0, 1]$ by $[0, \infty)$ as the following shows.

PROBLEM 8.6

For all non-negative integer n show that

$$\int_0^\infty x^n (\sin x^{1/4}) \exp(-x^{1/4}) \, dx = 0.$$

PROBLEM 8.7

Put $a_0 = 0$ and let $\{a_n\}_{n \geq 1}$ be a sequence of distinct positive numbers such that

$$\frac{1}{a_1} + \frac{1}{a_2} + \cdots + \frac{1}{a_n} + \cdots$$

diverges. Then show that $f \in C[0, 1]$ satisfies

$$\int_0^1 x^{a_n} f(x) \, dx = 0$$

for all $n \geq 0$ if and only if $f(x)$ vanishes everywhere on the interval $[0, 1]$.

This is a question posed by S. N. Bernstein (1880–1968) and solved by Müntz (1914) affirmatively. Carleman (1922) gave an elegant another proof using the theory of functions of a complex variable.

Solutions for Chapter 8

SOLUTION 8.1

For any positive ϵ it follows from the uniform continuity of f that there is a positive δ such that $|f(x) - f(y)| < \epsilon$ for any x and y in the interval $[0, 1]$ with $|x - y| < \delta$. Let M be the maximum of $|f(x)|$ on $[0, 1]$ and let $\{d_n\}$ be any monotone sequence of positive integers diverging to ∞ and satisfying $d_n = o(n)$ as $n \to \infty$. We divide the difference into two parts as follows:

$$B_n(f; x) - f(x) = \sum_{k=0}^{n} \left(f\left(\frac{k}{n}\right) - f(x) \right) \binom{n}{k} x^k (1 - x)^{n-k}$$

$$= S_0(x) + S_1(x),$$

where in $S_0(x)$ the summation runs through the values of $k \in [0, n]$ for which $|x - k/n| \leq d_n/n$ and in $S_1(x)$ the summation runs through the remaining values of k.

Now we consider all sufficiently large n satisfying $d_n/n < \delta$. For the sum $S_0(x)$, using $|f(x) - f(k/n)| < \epsilon$ we get

$$|S_0(x)| \leq \epsilon \sum_{k=0}^{n} \binom{n}{k} x^k (1 - x)^{n-k} = \epsilon.$$

On the other hand, for the sum $S_1(x)$, using $|j - nx| > d_n$ we obtain

$$|S_1(x)| < 2M \sum_{k=0}^{n} \left(\frac{j - nx}{d_n} \right)^2 \binom{n}{k} x^k (1 - x)^{n-k}$$

$$= \frac{2M}{d_n^2} (n^2 x^2 + nx(1 - x) - 2n^2 x^2 + n^2 x^2)$$

$$= \frac{2M}{d_n^2} nx(1 - x) \leq \frac{nM}{2d_n^2},$$

where we used the fact that $B_n(1; x) = 1$, $B_n(x; x) = x$ and

$$B_n(x^2; x) = x^2 + \frac{x(1 - x)}{n}.$$

We then take $d_n = [n^{2/3}]$ so that $d_n/\sqrt{n}$ diverges as $n \to \infty$. Hence we have $|S_0(x) + S_1(x)| < 2\epsilon$ for all sufficiently large n. $\square$

Remark. There is also an astonishing proof due to Bohman (1952) and Korovkin (1953), using only the linearity, monotonicity and the uniform convergence for $1, x, x^2$ among basic properties of B_n as an operator.

Let us explain this. For any $\epsilon > 0$ take δ and M as in the above proof. Putting

$$c = \sup_{|x-y| \geq \delta} \frac{|f(x) - f(y)|}{(x - y)^2},$$

we have

$$-\epsilon - c(x - y)^2 \leq f(x) - f(y) \leq \epsilon + c(x - y)^2$$

for all point (x, y) in the unit square $[0, 1]^2$. Applying the operator B_n to the function $\phi_y(x) = (x - y)^2$ with a parameter y, it follows from the linearity and monotonicity of B_n that

$$-\epsilon B_n(1; x) - c B_n(\phi_y; x) \leq B_n(f; x) - f(y) B_n(1; x)$$
$$\leq \epsilon B_n(1; x) + c B_n(\phi_y; x).$$

Let $\varphi_n(x)$ be the function obtained from $B_n(\phi_y; x)$ by carrying the substitution $y = x$. Since

$$B_n(\phi_y; x) = B_n(x^2; x) - 2y B_n(x; x) + y^2 B_n(1; x),$$

we have

$$\varphi_n(x) = (B_n(x^2; x) - x^2) - 2x(B_n(x; x) - x) + x^2(B_n(1; x) - 1),$$

which converges to 0 uniformly on the interval $[0, 1]$ as $n \to \infty$. Therefore

$$|B_n(f; x) - f(x)| \leq |B_n(f; x) - f(x) B_n(1; x)| + M|B_n(1; x) - 1|$$
$$\leq \epsilon + c|\varphi_n(x)| + (M + \epsilon)|B_n(1; x) - 1|,$$

which is less than 3ϵ uniformly in x for all sufficiently large n, as required.

The reason why we do not use the exact expressions of $B_n(1; x)$, $B_n(x; x)$ and $B_n(x^2; x)$ in the above proof, is to clarify the role of B_n as an operator; hence the proof is indeed valid for any linear operators having the monotonicity and the uniform convergence for $1, x, x^2$.

SOLUTION 8. 2

The Bernstein polynomial for f defined in **Problem 8.1** is

$$B_n(f; x) = \sum_{k=1}^{n-1} f\left(\frac{k}{n}\right)\binom{n}{k} x^k (1 - x)^{n-k}$$

since $f(0) = f(1) = 0$. If $n = p$ is a prime number, then the binomial coefficient $\binom{p}{k}$ is clearly a multiple of p for every integer k in $[1, p)$; hence

$$\widetilde{B}_p(f; x) = \sum_{k=1}^{p-1} \frac{1}{p} \binom{p}{k} \left[f\left(\frac{k}{p}\right) p \right] x^k (1-x)^{p-k}$$

is a polynomial with integral coefficients where $[x]$ denotes the integral part of x. Since we have

$$\left| B_p(f; x) - \widetilde{B}_p(f; x) \right| \leq \frac{1}{p} \sum_{k=1}^{p-1} \binom{p}{k} x^k (1-x)^{p-k} < \frac{1}{p},$$

it is clear that $\widetilde{B}_p(f; x)$ converges also to $f(x)$ uniformly as $p \to \infty$. □

SOLUTION 8.3

For any continuous periodic function with period 2π, we write $f(x) = f_+(x) + f_-(x)$ where

$$f_{\pm}(x) = \frac{f(x) \pm f(-x)}{2}$$

respectively. $f_{\pm}$ are also continuous periodic function with the same period satisfying $f_+(x) = f_+(-x)$ and $f_-(x) + f_-(-x) = 0$. Note that $f_-(k\pi) = 0$ for any integer k.

For any $\epsilon > 0$ we can take a continuous odd function $\phi(x)$ with period 2π such that $|f_-(x) - \phi(x)| < \epsilon$ for any $x \in \mathbb{R}$ and that $\phi(x)$ vanishes on every point of some small neighborhoods of the points $k\pi$. Since $x = \arccos y$ maps the interval $[-1, 1]$ onto $[0, \pi]$ homeomorphically, the functions

$$f_+(\arccos y) \quad \text{and} \quad \frac{\phi(\arccos y)}{\sin(\arccos y)}$$

are continuous on the interval $[-1, 1]$. Thus applying Weierstrass' approximation theorem to these functions, we can find certain polynomials $P(y)$ and $Q(y)$ satisfying

$$|f_+(x) - P(\cos x)| < \epsilon \quad \text{and} \quad |\phi(x) - (\sin x) Q(\cos x)| < \epsilon$$

for any $0 \leq x \leq \pi$. Moreover the inequalities hold for any $x \in \mathbb{R}$ by the evenness of $f_+(x)$ and $P(\cos x)$, by the oddness of $\phi(x)$ and $(\sin x) Q(\cos x)$ and, of course,

120   *Problems and Solutions in Real Analysis*

by the periodicity of these functions. We therefore obtain

$$|f(x) - P(\cos x) - (\sin x) Q(\cos x)|$$
$$\leq |f_+(x) - P(\cos x)| + |f_-(x) - \phi(x)|$$
$$+ |\phi(x) - (\sin x) Q(\cos x)|$$
$$< 3\epsilon$$

for any $x \in \mathbb{R}$. Finally it is easily verified that each $\cos^k x$ can be written as a linear combination of 1, $\cos x$, ..., $\cos kx$ and that each $(\sin x) \cos^k x$ can be expressed as a linear combination of $\sin x$, $\sin 2x$, ..., $\sin kx$. This completes the proof. □

Remark. This may be simpler than Achieser (1956)'s proof on p. 32 of his book.

SOLUTION 8.4

Since

$$B_n'(f;x) = \sum_{k=0}^{n} f\left(\frac{k}{n}\right)\binom{n}{k}\left(kx^{k-1}(1-x)^{n-k} - (n-k)x^k(1-x)^{n-k-1}\right)$$
$$= n \sum_{k=0}^{n-1}\left(f\left(\frac{k+1}{n}\right) - f\left(\frac{k}{n}\right)\right)\binom{n-1}{k}x^k(1-x)^{n-k-1},$$

it follows from the mean value theorem that

$$B_n'(f;x) = \sum_{k=0}^{n-1} f'\left(\frac{k+\xi_k}{n}\right)\binom{n-1}{k}x^k(1-x)^{n-k-1}$$

for some ξ_k in the interval $(0,1)$. Since for any $\epsilon > 0$ there exists an integer N such that $|f'(x) - f'(y)| < \epsilon$ for all x and y in $[0,1]$ with $|x-y| \leq 1/N$, we have

$$\left|B_n'(f;x) - B_{n-1}(f';x)\right|$$
$$\leq \sum_{k=0}^{n-1}\left|f'\left(\frac{k+\xi_k}{n}\right) - f'\left(\frac{k}{n}\right)\right|\binom{n-1}{k}x^k(1-x)^{n-k-1}$$
$$< \epsilon \sum_{k=0}^{n-1}\binom{n-1}{k}x^k(1-x)^{n-k-1} = \epsilon$$

for all integers $n > N$. □

SOLUTION 8.5

For any $\epsilon > 0$ there exists a polynomial $P(x)$ satisfying $|f(x) - P(x)| < \epsilon$ on the interval $[0, 1]$ by Weierstrass' approximation theorem. Letting M be the maximum of $|f(x)|$ on $[0, 1]$ we get

$$\int_0^1 f^2(x)\,dx = \int_0^1 (f(x) - P(x))\,f(x)\,dx + \int_0^1 P(x)f(x)\,dx$$

$$\leq \int_0^1 |f(x) - P(x)| \cdot |f(x)|\,dx < \epsilon M.$$

Since ϵ is arbitrary, we have $\int_0^1 f^2(x)\,dx = 0$; hence $f(x)$ vanishes everywhere. $\qquad\square$

SOLUTION 8.6

For brevity put

$$I_n = \int_0^\infty x^n e^{-x} \sin x\,dx \quad \text{and} \quad J_n = \int_0^\infty x^n e^{-x} \cos x\,dx$$

for any non-negative integer n. By the substitution $t = x^{1/4}$ the given integral in the problem is equal to $4I_{4n+3}$. By partial integration we easily get

$$\begin{cases} I_n = \dfrac{n}{2}\,(I_{n-1} + J_{n-1}), \\[2mm] J_n = \dfrac{n}{2}\,(J_{n-1} - I_{n-1}) \end{cases}$$

for any positive integer n. Solving these recursion formulae with the initial condition $I_0 = J_0 = 1/2$, we obtain $I_n = 0$ for any integer n satisfying $n \equiv 3 \pmod 4$. $\qquad\square$

REMARK. We also get $J_n = 0$ for any integer n satisfying $n \equiv 1 \pmod 4$. This means that

$$\int_0^\infty x^n \frac{(\cos x^{1/4})\exp(-x^{1/4})}{\sqrt{x}}\,dx = 0$$

for all non-negative integers n.

SOLUTION 8.7

Let m be any positive integer satisfying $m \neq a_n$ for all $n \geq 0$. We first consider the definite integral

$$I_n = \int_0^1 \left(x^m - c_0 - c_1 x^{a_1} - c_2 x^{a_2} - \cdots - c_n x^{a_n} \right)^2 dx$$

for any positive integer n. Obviously I_n is a polynomial in $c_0, c_1, ..., c_n$ of degrees 2 and attains its minimum I_n^* at some point $(s_0, s_1, ..., s_n) \in \mathbb{R}^{n+1}$, which is a unique solution of the system of $n + 1$ linear equations:

$$\frac{s_0}{a_0 + a_k + 1} + \frac{s_1}{a_1 + a_k + 1} + \cdots + \frac{s_n}{a_n + a_k + 1} = \frac{1}{m + a_k + 1} \tag{8.1}$$

for $0 \leq k \leq n$. Here the coefficient matrix

$$A = \left(\frac{1}{a_i + a_j + 1} \right)_{0 \leq i, j \leq n} \in M_{n+1}(\mathbb{R})$$

is symmetric and the determinant can be written explicitly as

$$\det A = \frac{\displaystyle\prod_{0 \leq i < j \leq n} (a_i - a_j)^2}{\displaystyle\prod_{0 \leq i, j \leq n} (a_i + a_j + 1)}$$

by using the Cauchy determinant. Now it follows from (8.1) that

$$I_n^* = \frac{1}{2m + 1} - \frac{s_0}{m + a_0 + 1} - \frac{s_1}{m + a_1 + 1} - \cdots - \frac{s_n}{m + a_n + 1}. \tag{8.2}$$

Combining (8.1) and (8.2) we thus have

$$\begin{pmatrix} & & 0 \\ A & & \vdots \\ & & 0 \\ a & & 1 \end{pmatrix} \begin{pmatrix} s_0 \\ \vdots \\ s_n \\ I_n^* \end{pmatrix} = \begin{pmatrix} {}^t a \\ \\ \frac{1}{2m + 1} \end{pmatrix}$$

where $a = ((m + a_0 + 1)^{-1}, ..., (m + a_n + 1)^{-1}) \in \mathbb{R}^{n+1}$. Therefore we get

$$I_n^* = \frac{\det B}{\det A}$$

by Cramer's rule where

$$B = \begin{pmatrix} A & {}^t a \\ a & \dfrac{1}{2m+1} \end{pmatrix} \in M_{n+2}(\mathbb{R})$$

is again symmetric and $\det B$ can be obtained from the Cauchy determinant by substituting a_{n+1} by m formally. We thus have

$$I_n^* = \frac{\det B}{\det A} = \frac{1}{2m+1} \prod_{k=0}^{n} \left(\frac{a_k - m}{a_k + m + 1} \right)^2 .$$

Since

$$\log \frac{|a_k - m|}{a_k + m + 1} = \log \left(1 - \frac{2m+1}{a_k + m + 1} \right)$$
$$\leq -\frac{2m+1}{a_k + m + 1} \leq -\frac{m}{a_k}$$

for any k satisfying $a_k > m$ and $\sum_{k=1}^{\infty} 1/a_k = \infty$ by the assumption, we infer that I_n^* converges to 0 as $n \to \infty$.

For any $\epsilon > 0$ we can take a sufficiently large n and $(c_0, c_1, ..., c_n) \in \mathbb{R}^{n+1}$ such that

$$\int_0^1 (x^m - c_0 - c_1 x^{a_1} - c_2 x^{a_2} - \cdots - c_n x^{a_n})^2 dx < \epsilon.$$

By the Cauchy-Schwarz inequality we obtain

$$\left(\int_0^1 x^m f(x) \, dx \right)^2 = \left(\int_0^1 (x^m - c_0 - c_1 x^{a_1} - c_2 x^{a_2} - \cdots - c_n x^{a_n}) f(x) \, dx \right)^2$$
$$\leq M \int_0^1 (x^m - c_0 - c_1 x^{a_1} - c_2 x^{a_2} - \cdots - c_n x^{a_n})^2 dx$$
$$< \epsilon M$$

where $M = \int_0^1 f^2(x) \, dx$. Since ϵ is arbitrary, $\int_0^1 x^m f(x) \, dx = 0$ for any positive integer m satisfying $m \neq a_n$ for all $n \geq 0$. Hence $\int_0^1 x^m f(x) \, dx = 0$ for all non-negative integers n and $f(x)$ vanishes everywhere on the interval $[0, 1]$ by **PROBLEM 8.5**.   □

Problems and Solutions in Real Analysis

Remark. For any positive integers n and m satisfying $a_k \neq m$ for all $k \geq 0$, define

$$d_{n,m} = \inf_{c_0,\ldots,c_n} \max_{0 \leq x \leq 1} \left| x^m - c_0 - c_1 x^{a_1} - \cdots - c_n x^{a_n} \right|$$

where the infimum ranges over all real numbers $c_0, \ldots, c_n$. The argument in the above proof can be used to show that $d_{n,m}$ converges to 0 as $n \to \infty$ for any fixed m. To see this, by using the Cauchy-Schwarz inequality, we have

$$\left| x^m - c_1 x^{a_1} - \cdots - c_n x^{a_n} \right|^2$$

$$= \left| \int_0^x \left(m t^{m-1} - a_1 c_1 t^{a_1 - 1} - \cdots - a_n c_n t^{a_n - 1} \right) dt \right|^2$$

$$\leq m^2 \int_0^1 \left(t^{m-1} - c_1' t^{a_1 - 1} - \cdots - c_n' t^{a_n - 1} \right)^2 dt$$

where $c_k' = c_k a_k / m$. We can now take n and $c_1', \ldots, c_n'$ suitably so that the right-hand side becomes arbitrarily small.

Chapter 9

Convex Functions

Some results in this chapter are due to J. L. W. V. Jensen (1859–1925) who introduced the notion of convex functions in Jensen (1906).

- A function $f(x)$ defined on an interval I is said to be *convex in the sense of Jensen* provided that

$$f\left(\frac{x+y}{2}\right) \le \frac{f(x) + f(y)}{2}$$

for any x and y in I. Note that $f(x)$ is not necessarily continuous on I.

- The set of all convex functions defined on I forms a positive cone; that is, if $f_1(x)$ and $f_2(x)$ are convex functions on I, then $c_1 f_1(x) + c_2 f_2(x)$ are also convex for any positive constants c_1 and c_2. Moreover $f(x) + a + bx$ is convex for any real constants a and b.

  For example, the function $|x|$ is clearly convex on $\mathbb{R}$; hence a piecewise linear function

$$\sum_{k=1}^{n} c_k |x - x_k|$$

is also convex on $\mathbb{R}$ for any positive constants $c_1, c_2, ..., c_n$.

- A function $g(x)$ is said to be *concave* if $-g(x)$ is convex.

- A positive function $f(x)$ is said to be *logarithmically convex* if $\log f(x)$ is convex.

PROBLEM 9.1

Suppose that $f(x)$ is convex on an interval I. Show that

$$f\left(\frac{x_1 + x_2 + \cdots + x_n}{n}\right) \le \frac{f(x_1) + f(x_2) + \cdots + f(x_n)}{n}$$

for arbitrary n points $x_1, x_2, \ldots, x_n$ in I.

PROBLEM 9.2

Show that $f(x)$ is convex on an interval I if and only if $e^{\lambda f(x)}$ is convex on I for any positive λ.

PROBLEM 9.3

Suppose that $f(x)$ is convex and bounded above on an open interval (a, b). Show then that $f(x)$ is continuous on (a, b).

PROBLEM 9.4

Suppose that $f(x)$ is convex and continuous on an interval I. Show that

$$f\left(\frac{\lambda_1 x_1 + \cdots + \lambda_n x_n}{\lambda_1 + \cdots + \lambda_n}\right) \le \frac{\lambda_1 f(x_1) + \cdots + \lambda_n f(x_n)}{\lambda_1 + \cdots + \lambda_n}$$

for any n points $x_1, \ldots, x_n$ in I and any positive numbers $\lambda_1, \ldots, \lambda_n$.

PROBLEM 9.5

Suppose that $g \in C[a, b]$ and $p(x)$ is a non-negative continuous function defined on $[a, b]$ satisfying $\sigma = \int_a^b p(x)\, dx > 0$. Let m and M be the minimum and maximum of the function $g(x)$ on $[a, b]$ respectively. Suppose further that f is a continuous convex function defined on $[m, M]$. Show then that

$$f\left(\frac{1}{\sigma}\int_a^b g(x)p(x)\, dx\right) \le \frac{1}{\sigma}\int_a^b f(g(x))p(x)\, dx.$$

PROBLEM 9.6

Show that any continuous convex function $f(x)$ on an open interval I possesses a finite derivative except for at most countable points.

Convex Functions   127

PROBLEM 9.7

Show that $f \in C[a, b]$ is convex if and only if

$$\frac{1}{t - s} \int_s^t f(x)\, dx \leq \frac{f(s) + f(t)}{2}$$

for any $s \neq t$ in $[a, b]$.

PROBLEM 9.8

Let I be a closed interval of the form either $[0, a]$ or $[0, \infty)$. Suppose that $f \in C(I)$ satisfies $f(0) = 0$. Show then that f is convex if and only if

$$\sum_{k=1}^{n} (-1)^{k-1} f(x_k) \geq f\left(\sum_{k=1}^{n} (-1)^{k-1} x_k \right)$$

for any integer $n \geq 2$ and any n points $x_1 \geq x_2 \geq \cdots \geq x_{n-1} \geq x_n$ in the interval I.

Wright (1954) gave a simple proof and noted that this is a special case of Theorem 108 in the book of Hardy, Littlewood and Pólya (1934).

PROBLEM 9.9

Let $s > -1$ be a real number. Suppose that $f \in C[0, \infty)$ is a convex function having the piecewise continuous derivative $f'(x)$ and satisfying $f(0) \geq 0$. Suppose further that $f'(0+)$ exists when $f(0) = 0$. Then show that

$$\int_0^\infty x^s \exp\left(-\frac{f(x)}{x} \right) dx \leq \int_0^\infty x^s \exp\left(-f'\left(\frac{x}{e}\right) \right) dx.$$

Prove moreover that the constant e in the denominator of the right-hand side cannot in general be replaced by any smaller number.

This is due to Carleson (1954). He used this inequality in the case $s = 0$ to show Carleman's inequality stated in **PROBLEM 2.3** in the following manner. Observe first that the given series can be arranged in decreasing order. He then defined the graph of $f(x)$ as the polygon whose vertices are the origin and the points

$$\left(n, \sum_{k=1}^{n} \log \frac{1}{a_k} \right)$$

for all positive integers n.

PROBLEM 9.10 —————————————————————————

Suppose that $f(x)$ is twice differentiable in an open interval I. Show then that $f(x)$ is convex if and only if $f''(x) \geq 0$ on I.

Note that we do not suppose the continuity of $f''(x)$.

PROBLEM 9.11 —————————————————————————

Suppose that $f \in C^2[0, \infty)$ is convex and bounded. Show that the improper integral

$$\int_0^\infty x f''(x)\, dx$$

converges.

Convex Functions   129

Solutions for Chapter 9

SOLUTION 9.1

By m times applications of convexity for f we easily get

$$f\left(\frac{x_1 + x_2 + \cdots + x_{2^m}}{2^m}\right) \le \frac{f(x_1) + f(x_2) + \cdots + f(x_{2^m})}{2^m}$$

for any 2^m points $x_1, ..., x_{2^m}$ in I. For any integer $n \ge 3$ we choose m satisfying $n < 2^m$. Now adding to $x_1, ..., x_n$ the new $2^m - n$ points

$$x_{n+1} = \cdots = x_{2^m} = \frac{x_1 + x_2 + \cdots + x_{2^m}}{n}$$

in I we have

$$2^m f(y) \le f(x_1) + \cdots + f(x_n) + (2^m - n) f\left(\frac{x_1 + x_2 + \cdots + x_n}{n}\right)$$

where

$$y = \frac{1}{2^m}\left(x_1 + x_2 + \cdots + x_n + (2^m - n)\frac{x_{11} + x_2 + \cdots + x_n}{n}\right)$$
$$= \frac{x_1 + x_2 + \cdots + x_n}{n};$$

namely

$$f\left(\frac{x_1 + x_2 + \cdots + x_n}{n}\right) \le \frac{f(x_1) + f(x_2) + \cdots + f(x_n)}{n}.$$

□

SOLUTION 9.2

Suppose first that $f(x)$ is convex on the interval I. Since $e^{\lambda x}$ is convex and monotone increasing on $\mathbb{R}$, we have

$$\exp\left(\lambda f\left(\frac{x_1 + x_2}{2}\right)\right) \le \exp\left(\lambda \frac{f(x_1) + f(x_2)}{2}\right) \le \frac{e^{\lambda f(x_1)} + e^{\lambda f(x_2)}}{2}$$

for any x_1 and x_2 in I. Hence $e^{\lambda f(x)}$ is convex on I.

Conversely suppose that $e^{\lambda f(x)}$ is convex on I for any positive λ. The asymptotic expansions as $\lambda \to 0+$ of both sides of the inequality

$$\exp\left(\lambda f\left(\frac{x_1 + x_2}{2}\right)\right) \le \frac{e^{\lambda f(x_1)} + e^{\lambda f(x_2)}}{2}$$

give

$$1 + \lambda f\left(\frac{x_1 + x_2}{2}\right) + O\left(\lambda^2\right) \le 1 + \lambda \frac{f(x_1) + f(x_2)}{2} + O\left(\lambda^2\right),$$

for any x_1 and x_2 in I, which implies the convexity of $f(x)$ on I. □

SOLUTION 9.3

Suppose that $f(x) < K$ for some constant K. For an arbitrary fixed y in the interval (a, b) and any positive integer n the points $y \pm n\delta$ belong to (a, b) for all sufficiently small positive δ. Of course δ depends on n. Applying the inequality described in **PROBLEM 9.1** to the n points $x_1 = y \pm n\delta, x_2 = \cdots = x_n = y$, we get

$$f(y \pm \delta) \le \frac{f(y \pm n\delta) + (n-1)f(y)}{n}$$

respectively. Hence we have

$$f(y + \delta) - f(y) \le \frac{f(y + n\delta) - f(y)}{n} \le \frac{K - f(y)}{n}$$

and

$$f(y) - f(y - \delta) \ge \frac{f(y) - f(y - n\delta)}{n} \ge \frac{f(y) - K}{n}.$$

Since $f(y) - f(y - \delta) \le f(y + \delta) - f(y)$ and n is arbitrary, these mean that $f(x)$ is continuous at the point y. □

SOLUTION 9.4

By the inequality in **PROBLEM 9.1** it is easily seen that

$$f\left(\frac{k_1 x_1 + \cdots + k_n x_n}{k_1 + \cdots + k_n}\right) \le \frac{k_1 f(x_1) + \cdots + k_n f(x_n)}{k_1 + \cdots + k_n}$$

for any points $x_1, ..., x_n$ in I and any positive integers $k_1, ..., k_n$. For any sufficiently large integer N we take $k_j = [\lambda_j N / (\lambda_1 + \cdots + \lambda_n)]$ for each $1 \le j \le n$. Since

$$\frac{k_j}{k_1 + \cdots + k_n} \to \frac{\lambda_j}{\lambda_1 + \cdots + \lambda_n}$$

as $N \to \infty$, the required inequality follows from the continuity of f. □

SOLUTION 9.5

We divide the interval $[a, b]$ into n equal parts and put

$$\lambda_k = \int_{t_{k-1}}^{t_k} p(x)\, dx \geq 0\,;$$

for any subinterval $[t_{k-1}, t_k]$ so that $\sigma = \sum_{k=1}^{n} \lambda_k$. It follows from the first mean value theorem that

$$\int_{t_{k-1}}^{t_k} g(x) p(x)\, dx = \lambda_k g(\xi_k)$$

for some ξ_k in (t_{k-1}, t_k). Applying the inequality in **PROBLEM 9.4** to n points $x_k = g(\xi_k)$ in $[m, M]$, we obtain

$$f\left(\frac{1}{\sigma} \int_a^b g(x) p(x)\, dx\right) = f\left(\frac{1}{\sigma} \sum_{k=1}^{n} \lambda_k g(\xi_k)\right)$$

$$\leq \frac{1}{\sigma} \sum_{k=1}^{n} \lambda_k f(g(\xi_k))$$

$$= \frac{b-a}{\sigma n} \sum_{k=1}^{n} f(g(\xi_k)) p(\eta_k)$$

for some η_k in (t_{k-1}, t_k). By the uniform continuity of $p(x)$, the difference between the expression on the right-hand side and one with $p(\eta_k)$ replaced by $p(\xi_k)$ is sufficiently small whenever n is sufficiently large. Therefore the right-hand side converges to

$$\frac{1}{\sigma} \int_a^b f(g(x)) p(x)\, dx$$

as $n \to \infty$.   □

SOLUTION 9.6

Let $f'_+(x)$ and $f'_-(x)$ be the right- and left-hand derivatives of f at x respectively. First we will prove that $f'_\pm(x)$ exist at every point x in the open interval I. For brevity let $\Delta(x, y)$ denote the difference quotient

$$\frac{f(x) - f(y)}{x - y}$$

for any $x \neq y$ in I. Let $x < y < z$ be arbitrary three points in I. Applying the inequality stated in **PROBLEM 9.4** with $\lambda_1 = z - y, x_1 = x$ and $\lambda_2 = y - x, x_2 = z$, we get $y = (\lambda_1 x_1 + \lambda_2 x_2)/(\lambda_1 + \lambda_2)$ and

$$f(y) \le \frac{z-y}{z-x} f(x) + \frac{y-x}{z-x} f(z).$$

Thus $\Delta(x, y) \le \Delta(y, z)$. Note that the above inequality can also be written as $\Delta(x, y) \le \Delta(x, z)$ or as $\Delta(x, z) \le \Delta(y, z)$. In particular, the quotients $\Delta(x - h, x)$ and $\Delta(x, x + h)$ are monotone increasing with respect to $h > 0$ and $\Delta(x - h, x) \le \Delta(x, x + h)$ holds for any $h > 0$ satisfying $x \pm h \in I$. This means that both $f'_+(x)$ and $f'_-(x)$ certainly exist and satisfy $f'_-(x) \le f'_+(x)$.

Next for any points $x < y$ in I we take a sufficiently small $h > 0$ satisfying $x + h < y - h$. Then

$$\Delta(x, x + h) \le \Delta(x + h, y - h) \le \Delta(y - h, y);$$

hence, letting h to $0+$ we obtain $f'_+(x) \le f'_-(y)$. Therefore if $f(x)$ does not possess a finite differential coefficient at x_0, then it follows that $f'_-(x_0) < f'_+(x_0)$. Moreover if we assign the point x_0 to the open interval $(f'_-(x_0), f'_+(x_0))$, then such intervals are disjoint each other. Therefore we can enumerate all such open intervals by labeling them, for example, in such a way that we count ones contained in $(-n, n)$ and having the length $> 1/n$ for each positive integer n. $\qquad\square$

SOLUTION 9.7

First assume that a continuous function $f(x)$ is convex on the interval $[a, b]$. We divide the subinterval $[s, t]$ into n equal parts and put

$$x_k = \frac{n-k}{n} s + \frac{k}{n} t$$

for $0 \le k \le n$. It follows from the inequality in PROBLEM 9.1 that

$$f(x_k) \le \frac{n-k}{n} f(s) + \frac{k}{n} f(t).$$

Therefore we have

$$\frac{1}{t-s} \int_s^t f(x)\,dx = \lim_{n\to\infty} \frac{1}{n} \sum_{k=0}^n f(x_k)$$

$$\le \limsup_{n\to\infty} \frac{n(n+1)}{2n^2} (f(s) + f(t)) = \frac{f(s) + f(t)}{2}.$$

Conversely assume that

$$\frac{1}{t-s} \int_s^t f(x)\,dx \le \frac{f(s) + f(t)}{2}$$

for any $s \neq t$ in $[a, b]$. Suppose, on the contrary, that there are two points $s < t$ in the interval $[a, b]$ satisfying

$$f\left(\frac{s+t}{2}\right) > \frac{f(s) + f(t)}{2}.$$

Then we consider the set

$$E = \left\{ x \in [s, t]; \ f(x) > f(s) + \frac{f(t) - f(s)}{t - s} (x - s) \right\}.$$

The set E is clearly open by the continuity of f and $E \neq \emptyset$ since it contains the point $(s + t)/2$. Note that E is the set of points on the interval $[s, t]$ at which the graph of $f(x)$ is situated in the upper side of the straight line through the two points $(s, f(s))$ and $(t, f(t))$. Let (u, v) be the connected component of E containing the point $(s + t)/2$. Since the end points $(u, f(u))$ and $(v, f(v))$ must be on that line, we have

$$\begin{cases} f(u) = f(s) + \dfrac{f(t) - f(s)}{t - s} (u - s), \\[2mm] f(v) = f(s) + \dfrac{f(t) - f(s)}{t - s} (v - s), \end{cases}$$

which imply that

$$\frac{f(u) + f(v)}{2} = f(s) + \frac{f(t) - f(s)}{t - s} \left(\frac{u + v}{2} - s\right).$$

Therefore

$$\frac{1}{v - u} \int_u^v f(x)\,dx > \frac{1}{v - u} \int_u^v \left(f(s) + \frac{f(t) - f(s)}{t - s} (x - s)\right) dx$$

$$= f(s) + \frac{f(t) - f(s)}{t - s} \left(\frac{u + v}{2} - s\right)$$

$$= \frac{f(u) + f(v)}{2},$$

contrary to the assumption. $\qquad\qquad\qquad\qquad\qquad\qquad\qquad\qquad\quad\square$

$\boxed{\text{SOLUTION } 9.8}$

Suppose first that f is convex on the closed interval I. Let $x_1 > x_2 > x_3$ be arbitrary three points in I and define a positive number λ with $x_2 = \lambda x_1 + (1 - \lambda)x_3$. Since f is convex, it follows from **PROBLEM 9.4** that

$$f(x_2) = f(\lambda x_1 + (1 - \lambda)x_3) \leq \lambda f(x_1) + (1 - \lambda)f(x_3)$$

and

$$f(x_1 - x_2 + x_3) = f((1 - \lambda)x_1 + \lambda x_3) \leq (1 - \lambda)f(x_1) + \lambda f(x_3);$$

therefore we have

$$f(x_2) + f(x_1 - x_2 + x_3) \leq f(x_1) + f(x_3),$$

which is also valid for any $x_1 \geq x_2 \geq x_3$ in I by the continuity of f. If we take $x_3 = 0$, then clearly $f(x_1 - x_2) \leq f(x_1) - f(x_2)$ by virtue of $f(0) = 0$. We thus have the inequality in the problem for the cases $n = 2$ and 3. Suppose now the inequality holds for $n = m \geq 2$. Then for any $m + 2$ points $x_1 \geq x_2 \geq \cdots \geq x_{m+2}$ in the interval I,

$$\sum_{k=1}^{m+2} (-1)^{k-1} f(x_k) \geq f(x_1) - f(x_2) + f\left(\sum_{k=3}^{m+2} (-1)^{k-1} x_k\right)$$

$$\geq f\left(\sum_{k=1}^{m+2} (-1)^{k-1} x_k\right),$$

which means that the inequality holds for $n = m + 2$; therefore for every $n \geq 2$. Conversely suppose the case $n = 3$:

$$f(x_2) + f(x_1 - x_2 + x_3) \leq f(x_1) + f(x_3)$$

for any points $x_1 \geq x_2 \geq x_3$ in I. By taking $x_2 = (x_1 + x_3)/2$ we get

$$f\left(\frac{x_1 + x_3}{2}\right) \leq \frac{f(x_1) + f(x_3)}{2};$$

hence f is convex on I.   □

SOLUTION 9.9

As is shown in SOLUTION 9.6, the difference quotient

$$\Delta(x + h, x) = \frac{f(x + h) - f(x)}{h}$$

is monotone increasing for $h > 0$; therefore we have $\Delta(\alpha x, x) \geq \Delta(x + h, x)$ for any real numbers $\alpha > 1$ and $x > 0$ if $(\alpha - 1)x \geq h > 0$ is fulfilled. Letting $h \to 0+$, we get $\Delta(\alpha x, x) \geq f'(x)$ if the derivative exists; in other words,

$$f(\alpha x) \geq f(x) + (\alpha - 1)x f'(x).$$

For brevity put

$$F(x) = x^s \exp\left(-\frac{f(x)}{x}\right) \quad \text{and} \quad G(x) = x^s \exp\left(-f'(x)\right).$$

Note that the improper integral

$$\int_0^\infty F(x)\,dx$$

converges at $x = 0$ for any $s > -1$ when $f(0) > 0$ or when $f(0) = 0$ and $f'(0+)$ exists, since $f'(0+) \le f(x)/x$ in the latter case. The convergence at $x = \infty$ also follows. By the substitution $x = \alpha t$ we have

$$\int_0^L F(x)\,dx = \alpha^{s+1} \int_0^{L/\alpha} t^s \exp\left(-\frac{f(\alpha t)}{\alpha t}\right) dt$$

$$\le \alpha^{s+1} \int_0^{L/\alpha} F^{1/\alpha}(t) G^{(\alpha-1)/\alpha}(t)\,dt$$

for any $L > 0$. Applying Hölder's inequality, the right-hand side is less than or equal to

$$\alpha^{s+1} \left(\int_0^{L/\alpha} F(x)\,dx\right)^{1/\alpha} \left(\int_0^{L/\alpha} G(x)\,dx\right)^{(\alpha-1)/\alpha}$$

and replacing L/α by L we have

$$\int_0^L F(x)\,dx < \phi^{s+1}(\alpha) \int_0^L G(x)\,dx$$

where $\phi(\alpha) = \alpha^{\alpha/(\alpha-1)}$ is strictly monotone increasing on the interval $(1, \infty)$. By letting $\alpha \to 1+$ we get

$$\int_0^L F(x)\,dx \le e^{s+1} \int_0^L G(x)\,dx$$

$$= \int_0^{eL} t^s \exp\left(-f'\left(\frac{t}{e}\right)\right) dt.$$

The desired inequality follows by letting $L \to \infty$.

To see that the constant e is best possible, we take $f(x) = x^\beta$ for any $\beta > 1$. Then it is not hard to see that

$$\int_0^\infty F(x)\,dx = \frac{1}{\beta - 1}\,\Gamma\left(\frac{s+1}{\beta-1}\right)$$

and

$$\int_0^\infty G(x)\,dx = \frac{1}{(\beta-1)\beta^{(s+1)/(\beta-1)}}\,\Gamma\left(\frac{s+1}{\beta-1}\right)$$

where $\Gamma(s)$ is the Gamma function. Hence the ratio of the two integrals converges to e^{s+1} as $\beta \to 1+$. $\square$

REMARK. Carleson (1954) obtained his inequality explained in this problem under the condition $f(0) = 0$. He also claimed that 'the sign of equality is excluded, but we shall not insist on this detail'.

$\boxed{\text{SOLUTION 9.10}}$

Suppose that $f(x)$ is convex on I. We have already seen in the proof of **PROBLEM 9.6** that $f'_+(x) \le f'_-(y)$ for any points $x < y$ in I. Since $f(x)$ is twice differentiable, the derivative $f'(x)$ is monotone increasing on I, whence $f''(x) \ge 0$.

Conversely suppose that $f''(x) \ge 0$. Let x and y be any points in I. By Taylor's formula centered at $(x+y)/2$ we obtain

$$f(x) = f\left(\frac{x+y}{2}\right) + f'\left(\frac{x+y}{2}\right)\frac{x-y}{2} + f''(c)\frac{(x-y)^2}{8}$$

and

$$f(y) = f\left(\frac{x+y}{2}\right) + f'\left(\frac{x+y}{2}\right)\frac{y-x}{2} + f''(c')\frac{(x-y)^2}{8}$$

for some c and c'. Adding the two equalities we get

$$f(x) + f(y) = 2f\left(\frac{x+y}{2}\right) + (f''(c) + f''(c'))\frac{(x-y)^2}{8}$$

$$\ge 2f\left(\frac{x+y}{2}\right).$$

Hence $f(x)$ is convex on I. $\square$

$\boxed{\text{SOLUTION 9.11}}$

Suppose $\delta = f'(x_0) > 0$ for some $x_0 > 0$. Since $f'(x)$ is monotone increasing, we have $f'(x) \ge \delta$ for any $x \ge x_0$ and hence

$$f(x) = f(x_0) + \int_{x_0}^x f'(t)\,dt$$

$$\ge f(x_0) + \delta(x - x_0),$$

contrary to the assumption that f is bounded. Thus $f'(x) \le 0$ for any $x > 0$ and it follows that $f(x)$ is monotone decreasing and converges to some λ as $x \to \infty$. Therefore we have

$$\lambda - f(x) = \int_x^\infty f'(t)\, dt$$

and, in particular, $f'(x)$ converges to 0 as $x \to \infty$. By the Cauchy criterion we have, for any $\epsilon > 0$,

$$0 < -\int_\alpha^\beta f'(t)\, dt < \epsilon$$

for all sufficiently large α and $\beta > \alpha$. Since $f'(x)$ is non-positive and monotone increasing, we obtain $(\beta - \alpha)|f'(\beta)| < \epsilon$; therefore

$$\beta |f'(\beta)| < \epsilon + \alpha |f'(\beta)|.$$

The right-hand side can be smaller than 2ϵ if we take β sufficiently large. Hence $xf'(x)$ converges to 0 as $x \to \infty$. Thus

$$\int_0^x t f''(t)\, dt = x f'(x) - f(x) + f(0)$$

converges to $f(0) - \lambda$ as $x \to \infty$. $\square$

Chapter 10

Various proofs of $\zeta(2) = \pi^2/6$

The standard way to evaluate the sum

$$\zeta(2) = \sum_{n=1}^{\infty} \frac{1}{n^2}$$

may be the usage of the partial fraction expansion for $\cot \pi z$:

$$\pi \cot \pi z = \frac{1}{z} + \sum_{n=1}^{\infty} \frac{2z}{z^2 - n^2}.$$

This method has an advantage of evaluating all $\zeta(2n)$ at the same time, but one needs certain justification for representations of meromorphic functions by partial fractions or factorizations, as is described in Ahlfors (1966) in detail. See also Kanemitsu and Tsukada (2007), where it is shown that the partial fraction expansion for the cotangent function is a form of the functional equation for the Riemann zeta function.

There is also a way of finding the sum $\zeta(2)$ using the Fourier series for suitable periodic continuous functions. For example, the following trigonometric series appeared in **PROBLEM 7.11**:

$$\sum_{n=1}^{\infty} \frac{\sin n\theta}{n},$$

which converges boundedly in the interval $(0, 2\pi)$ to $(\pi - \theta)/2$. Hence, integrating by parts we get

$$\sum_{n=1}^{\infty} \frac{1 - (-1)^n}{n^2} = \int_0^\pi \frac{\pi - \theta}{2} \, d\theta = \frac{\pi^2}{4},$$

which implies $\zeta(2) = \pi^2/6$. But one needs some justification for good convergence of the Fourier series to apply partial integration. In general, using the

trigonometric series of the Bernoulli polynomials one can obtain the closed form
of the sum $\zeta(2n)$, as is described in Dieudonné (1971) in detail.

Using hypergeometric series Choi and Rathie (1997) gave an evaluation for
$\zeta(2)$. See also Choi, Rathie and Srivastava (1999).

In all the problems below we understand that we are to evaluate the series
$\sum_{n=1}^{\infty} n^{-2}$. Various easier proofs for $\zeta(2) = \pi^2/6$ are collected so that the reader
will enjoy them.

PROBLEM 10.1

It follows from the Gregory-Leibniz series

$$\sum_{n=0}^{\infty} \frac{(-1)^n}{2n+1} = \frac{\pi}{4}$$

that the sequence

$$a_n = \sum_{k=-n}^{n} \frac{(-1)^k}{2k+1}$$

converges to $\pi/2$ as $n \to \infty$. Then square the a_n.

This is due to J. M. Borwein and P. B. Borwein (1987). The similar method
to evaluate $\zeta(2)$ from the Gregory-Leibniz series was already found by Denquin
(1912) and by Estermann (1947).

PROBLEM 10.2

The reciprocal of the function $\sin\theta$ is denoted by $\operatorname{cosec}\theta$. Use

$$\frac{1}{\theta^2} < \operatorname{cosec}^2\theta < 1 + \frac{1}{\theta^2}$$

for $0 < \theta < \pi/2$ and the formula

$$\operatorname{cosec}^2\theta = \frac{1}{4}\left(\operatorname{cosec}^2\frac{\theta}{2} + \operatorname{cosec}^2\frac{\theta+\pi}{2}\right).$$

This is due to Hofbauer (2002).

Various proofs of $\zeta(2) = \pi^2/6$ 141

PROBLEM 10.3

The reciprocal of the function of $\tan\theta$ *is denoted by* $\cot\theta$. *Use*

$$\cot^2\theta < \frac{1}{\theta^2} < 1 + \cot^2\theta$$

for $0 < \theta < \pi/2$ *and the formula*

$$\cot^2\frac{\pi}{2n+1} + \cdots + \cot^2\frac{n\pi}{2n+1} = \frac{n}{3}(2n-1).$$

This is due to A. M. Yaglom and I. M. Yaglom (1953). The same argument can be found in Holme (1970) and in Papadimitriou (1973). A similar but a little bit complicated proof was given by Kortram (1996). The same method was also discussed in Arratia (1999).

PROBLEM 10.4

Multiply the following formula by θ *and integrate from* 0 *to* $\pi/2$:

$$\frac{1}{2} + \cos 2\theta + \cos 4\theta + \cdots + \cos 2n\theta = \frac{\sin(2n+1)\theta}{2\sin\theta}.$$

This is due to Giesy (1972). A similar proof was given by Stark (1969), who multiplied by θ and integrated from 0 to $\pi/2$ the so-called Fejér kernel

$$\frac{\sin^2(n+1)\theta}{2(n+1)\sin^2\theta} = \frac{1}{2} + \sum_{k=1}^{n}\left(1 - \frac{k}{n+1}\right)\cos 2k\theta.$$

PROBLEM 10.5

Carry out the substitution $x = \sin\theta$ *in the Taylor series*

$$\arcsin x = x + \sum_{n=1}^{\infty}\frac{1\cdot 3\cdot 5\cdots(2n-1)}{2\cdot 4\cdot 6\cdots(2n)}\cdot\frac{x^{2n+1}}{2n+1}$$

valid for $|x| \le 1$ *and use the formula*

$$\int_0^{\pi/2}\sin^{2n+1}\theta\,d\theta = \frac{2\cdot 4\cdots(2n)}{3\cdot 5\cdots(2n+1)}.$$

This is due to Choe (1987), who used substantially the value $(\arcsin 1)^2$. However this is very close to Euler's proof as described in Ayoub (1974) and Kimble (1987) reproduced it without words.

PROBLEM 10.6

Show first that

$$\zeta(2) = 3 \sum_{n=1}^{\infty} \frac{(n-1)!^2}{(2n)!}.$$

To evaluate the value of the series on the right-hand side, use the Taylor series

$$\arcsin^2 x = \sum_{n=1}^{\infty} \frac{2^{2n-1}(n-1)!^2}{(2n)!} x^{2n}.$$

This is due to Knopp and Schur (1918).

PROBLEM 10.7

Putting

$$I_n = \int_0^{\pi/2} \cos^{2n}\theta \, d\theta \quad and \quad J_n = \int_0^{\pi/2} \theta^2 \cos^{2n}\theta \, d\theta$$

for any non-negative integer n, show the recursion formula

$$I_n = n(2n-1)J_{n-1} - 2n^2 J_n.$$

Use then the formula

$$\int_0^{\pi/2} \cos^{2n}\theta \, d\theta = \frac{1 \cdot 3 \cdots (2n-1)}{2 \cdot 4 \cdots (2n)} \cdot \frac{\pi}{2}.$$

This is due to Matsuoka (1961). See the comment after **PROBLEM 18.6**.

PROBLEM 10.8

Carry out the substitution

$$2\cos\theta = e^{\theta i} + e^{-\theta i}$$

in the improper integral

$$\int_0^{\pi/2} \log(2\cos\theta) \, d\theta.$$

This is due to Russell (1991).

Various proofs of $\zeta(2) = \pi^2/6$ 143

PROBLEM 10.9

For any fixed x in the interval $(0, 1]$ carry out the substitution

$$\cos\theta = y + \frac{x}{2}(y^2 - 1)$$

in the improper repeated integral

$$\int_0^1 dx \int_{-1}^1 \frac{dy}{1 + xy}.$$

This is due to Goldscheider (1913) who answered the question posed by Stäckel (1913) as Aufgabe 208.

PROBLEM 10.10

Carry out the affine transformation

$$\begin{cases} u = (x + y)/\sqrt{2} \\ v = (y - x)/\sqrt{2} \end{cases}$$

in the improper double integral

$$\zeta(2) = \iint_S \frac{dx\,dy}{1 - xy},$$

where S is the unit square $[0, 1) \times [0, 1)$, in order to show that

$$\zeta(2) = 4 \int_0^{1/\sqrt{2}} \arctan\frac{t}{\sqrt{2 - t^2}} \frac{dt}{\sqrt{2 - t^2}}$$

$$+ 4 \int_{1/\sqrt{2}}^{\sqrt{2}} \arctan\frac{\sqrt{2} - t}{\sqrt{2 - t^2}} \frac{dt}{\sqrt{2 - t^2}}.$$

Then substitute $t = \sqrt{2}\sin\theta$ and $t = \sqrt{2}\cos 2\theta$ in the above integrals respectively.

This is due to Apostol (1983).

PROBLEM 10.11

Carry out the transformation

$$\begin{cases} x = \sin\theta/\cos\varphi \\ y = \sin\varphi/\cos\theta \end{cases}$$

on the triangular region $\{(\theta,\varphi);\, \theta,\varphi > 0,\ \theta+\varphi < \pi/2\}$ in the improper double integral

$$\frac{3}{4}\zeta(2) = \iint\limits_{S} \frac{dx\,dy}{1 - x^2 y^2},$$

where S is the unit square $[0,1) \times [0,1)$.

According to Kalman (1993) this proof was given in a lecture by Don B. Zagier in 1989, who mentioned that it was shown to him by a colleague who had learned of it through the grapevine. Elkies (2003) has reported that this proof, as well as the higher dimensional generalization, is due to Calabi and that the only paper containing the proof is Beukers, Calabi and Kolk (1993). See **PROBLEM 11.4**.

PROBLEM 10.12

Interchange the order of integration of the improper repeated integral

$$\int_0^\infty \frac{x\,dx}{1+x^2} \int_0^1 \frac{dy}{1+x^2 y^2}.$$

This is due to Harper (2003).

PROBLEM 10.13

The power series

$$D(z) = \sum_{n=1}^{\infty} \frac{z^n}{n^2}$$

converges absolutely on $|z| \leq 1$ with the radius of convergence 1 and satisfies the relations $D(1) = \zeta(2)$ and $D(-1) = -\zeta(2)/2$. We have

$$D(z) = -\int_0^z \frac{\log(1-z)}{z} \, dz,$$

from which $D(z)$ has an analytic continuation to the whole complex plane except for the half line $[1, \infty)$ by taking the principal branch for the logarithm $\log(1-z)$. Show then the functional equation

$$D\left(-\frac{1}{z}\right) + D(-z) = 2D(-1) - \frac{1}{2}\log^2 z$$

and consider the limit as $z \to -1$ in the upper half plane.

The function $D(z)$ is called the dilogarithm. This is taken from Levin's book (1981).

Solutions for Chapter 10

SOLUTION 10.1

Put

$$b_n = \sum_{k=-n}^{n} \frac{1}{(2k+1)^2}$$

for brevity. Then we have

$$
\begin{aligned}
a_n^2 - b_n &= \sum_{-n \le k \ne m \le n} \frac{(-1)^{k+m}}{(2k+1)(2m+1)} \\
&= \sum_{-n \le k \ne m \le n} \frac{(-1)^{k+m}}{2(m-k)} \left(\frac{1}{2k+1} - \frac{1}{2m+1} \right) \\
&= \sum_{-n \le k \ne m \le n} \frac{(-1)^{k+m}}{(m-k)(2k+1)}.
\end{aligned}
$$

The last expression can be written as

$$\sum_{k=-n}^{n} \frac{(-1)^k}{2k+1} c_{k,n}$$

where the summation

$$c_{k,n} = \sum \frac{(-1)^m}{m-k}$$

runs through the integral values of m in $[-n, n]$ except for the value of k. The sum $c_{k,n}$ contains several terms to be canceled. Indeed,

$$c_{k,n} = (-1)^{k+1} \sum_{\ell=n-k+1}^{k+n} \frac{(-1)^\ell}{\ell}$$

for positive k. It is also clear that $c_{0,n} = 0$ and $c_{-k,n} = -c_{k,n}$; so

$$|c_{k,n}| \le \frac{1}{n+|k|+1}.$$

Hence

$$\left| a_n^2 - b_n \right| \leq \sum_{k=-n}^{n} \frac{1}{|2k+1|(n-|k|+1)}$$
$$< \frac{1}{n} + \frac{6}{2n+1}\left(1 + \frac{1}{2} + \cdots + \frac{1}{2n+1}\right).$$

The right-hand side converges to 0 as $n \to \infty$ and thus b_n converges to $\pi^2/4$, which is equal to $3\zeta(2)/2$. $\square$

SOLUTION 10.2

Applying repeatedly the formula described in the problem, starting with $1 = \cosec^2(\pi/2)$, we get

$$1 = \frac{1}{4}\left(\cosec^2 \frac{\pi}{4} + \cosec^2 \frac{3\pi}{4}\right)$$
$$= \frac{1}{16}\left(\cosec^2 \frac{\pi}{8} + \cosec^2 \frac{3\pi}{8} + \cosec^2 \frac{5\pi}{8} + \cosec^2 \frac{7\pi}{8}\right)$$
$$\vdots$$
$$= \frac{1}{4^n} \sum_{k=0}^{2^n-1} \cosec^2 \frac{2k+1}{2^{n+1}}\pi.$$

On the other hand, we have

$$\theta > \sin\theta > \theta - \frac{\theta^3}{6} > \frac{\theta}{\sqrt{1+\theta^2}}$$

for $0 < \theta < \pi/2$, which implies the inequality on cosec stated in the problem. Using that inequality, we have

$$4^{n+1}s_n < 2^{2n-1} < 2^n + 4^{n+1}s_n$$

where

$$s_n = \frac{1}{\pi^2} \sum_{k=0}^{2^{n-1}-1} \frac{1}{(2k+1)^2}.$$

Dividing both sides by 4^{n+1} and letting n to ∞ we find that the sequence s_n converges to $1/8$, which is equal to $\frac{3}{4\pi^2}\zeta(2)$. $\square$

SOLUTION 10.3

Putting

$$\phi(\theta) = \sin\theta - \theta\cos\theta$$

for $0 < \theta < \pi/2$, we have $\phi(0+) = 0$, $\phi'(\theta) = \theta\sin\theta > 0$; so $\theta < \tan\theta$ and hence

$$\cot^2\theta < \frac{1}{\theta^2} < 1 + \cot^2\theta.$$

Let

$$\xi_k = \cot^2\frac{k\pi}{2n+1}$$

for positive integer k. Then

$$\frac{c_n}{(2n+1)^2} < \sum_{k=1}^{n}\frac{1}{(k\pi)^2} < \frac{n}{(2n+1)^2} + \frac{c_n}{(2n+1)^2}$$

where $c_n = \xi_1 + \xi_2 + \cdots + \xi_n$. Thus it suffices to show that c_n/n^2 converges to $2/3$ as $n \to \infty$.

To see this we introduce a polynomial of degree n vanishing at ξ_k for $1 \le k \le n$. We now have

$$\sin(2n+1)\theta = \Im\, e^{(2n+1)\theta i} = \Im\sum_{k=0}^{2n+1} i^k\binom{2n+1}{k}\cos^{2n+1-k}\theta\,\sin^k\theta$$

$$= \sum_{\ell=0}^{n}(-1)^\ell\binom{2n+1}{2\ell+1}\cos^{2n-2\ell}\theta\,\sin^{2\ell+1}\theta.$$

Dividing both sides by $\sin^{2n+1}\theta$ we see that

$$\frac{\sin(2n+1)\theta}{\sin^{2n+1}\theta} = \sum_{\ell=0}^{n}(-1)^\ell\binom{2n+1}{2\ell+1}\cot^{2n-2\ell}\theta.$$

The right-hand side is a polynomial of $\cot^2\theta$ of degree n, which we denote by $Q(\cot^2\theta)$. Solving the equation $\sin(2n+1)\theta = 0$ with $\sin\theta \ne 0$, we see that ξ_1, ..., ξ_n are n real simple zeros of $Q(x)$. Since

$$Q(x) = (2n+1)x^n - \frac{n}{3}(4n^2-1)x^{n-1} + \cdots,$$

we have $c_n = n(2n-1)/3$, whence $c_n/n^2 \to 2/3$ as $n \to \infty$, as required.   □

SOLUTION 10.4

The formula in the problem will be easily shown in the same way as in the proof of **PROBLEM 1.7**. Using

$$\int_0^{\pi/2} \theta \cos 2k\theta \, d\theta = -\frac{1}{2k} \int_0^{\pi/2} \sin 2k\theta \, d\theta = \frac{(-1)^k - 1}{4k^2}$$

it can be seen that the resulting value of the left-hand side in the problem for $n = 2m + 1$ is equal to

$$\frac{\pi^2}{16} - \frac{1}{2} \left(\frac{1}{1^2} + \frac{1}{3^2} + \cdots + \frac{1}{(2m+1)^2} \right),$$

which converges to $\pi^2/16 - 3\zeta(2)/8$ as $m \to \infty$. On the other hand, it follows from **PROBLEM 5.1** that

$$\frac{1}{2} \int_0^{\pi/2} \frac{\theta}{\sin \theta} \sin(4m+3)\theta \, d\theta$$

converges to 0 as $m \to \infty$. Note that we multiplied θ so that the function $\theta/\sin\theta$ is continuous on the interval $[0, \pi/2]$; otherwise the function $1/\sin\theta$ is not integrable on that interval. $\qquad\square$

SOLUTION 10.5

The Taylor series of the function $(1 - x^2)^{-1/2}$ about $x = 0$ is

$$(1 - x^2)^{-1/2} = 1 + \frac{1}{2}x^2 + \frac{1\cdot 3}{2\cdot 4}x^4 + \frac{1\cdot 3\cdot 5}{2\cdot 4\cdot 6}x^6 + \cdots,$$

whose radius of convergence is equal to 1. The termwise integration yields the Taylor series for $\arcsin x$ as follows:

$$\arcsin x = x + \sum_{n=1}^{\infty} \frac{1\cdot 3\cdot 5 \cdots (2n-1)}{2\cdot 4\cdot 6 \cdots (2n)} \cdot \frac{x^{2n+1}}{2n+1},$$

which is valid for $|x| < 1$. Moreover this power series converges uniformly on the interval $[-1, 1]$ since Stirling's approximation implies that

$$\frac{1\cdot 3\cdot 5 \cdots (2n-1)}{2\cdot 4\cdot 6 \cdots (2n)} = \binom{2n}{n} 4^{-n} = O\left(\frac{1}{\sqrt{n}} \right)$$

as $n \to \infty$. Thus by the substitution $x = \sin \theta$ in the Taylor series of arcsin x and integrating from 0 to $\pi/2$ we get

$$\int_0^{\pi/2} \theta \, d\theta = 1 + \sum_{n=1}^{\infty} \frac{1}{(2n+1)^2},$$

which implies that $\pi^2/8 = 3\zeta(2)/4$.   □

$\boxed{\text{SOLUTION } 10.6}$

First we put

$$\sigma_m = \sum_{n \geq m} \frac{m!}{n^2(n+1)\cdots(n+m)}$$

for any positive integer m. The first difference of this sequence is

$$\sigma_m - \sigma_{m+1} = \frac{(m-1)!^2}{(2m)!} + \sum_{n>m} \left(\frac{m!}{n^2(n+1)\cdots(n+m)} - \frac{(m+1)!}{n^2(n+1)\cdots(n+m+1)} \right).$$

It is easily seen that the infinite sum on the right-hand side can be written as

$$\sum_{n>m} \frac{m!}{n(n+1)\cdots(n+m+1)},$$

which is transformed into

$$\frac{1}{m+1} \sum_{n>m} \sum_{k=0}^{m+1} (-1)^k \binom{m+1}{k} \frac{1}{n+k} = \frac{1}{m+1} \int_0^1 t^m (1-t)^m \, dt$$

$$= \frac{m!^2}{(m+1)(2m+1)!}.$$

On the other hand, since

$$\sigma_1 = \sum_{n=1}^{\infty} \frac{1}{n^2(n+1)} = \sum_{n=1}^{\infty} \frac{1}{n} \left(\frac{1}{n} - \frac{1}{n+1} \right) = \zeta(2) - 1,$$

we obtain

$$\zeta(2) - \sigma_k = 1 + \sigma_1 - \sigma_k = 1 + \sum_{m=1}^{k-1} (\sigma_m - \sigma_{m+1})$$

$$= 1 + \sum_{m=1}^{k-1} \left(\frac{(m-1)!^2}{(2m)!} + \frac{m!^2}{(m+1)(2m+1)!} \right),$$

which is equal to

$$\sum_{m=1}^{k-1} \frac{(m-1)!^2}{(2m)!} + 2\sum_{m=1}^{k} \frac{(m-1)!^2}{(2m)!}$$

for any positive integer k. Therefore, noting that

$$0 < \sigma_k < \int_0^1 t^{k-1}(1-t)^{k-1}\, dt$$

and the right-hand side converges to 0 as $k \to \infty$, we have the formula

$$\zeta(2) = 3\sum_{n=1}^{\infty} \frac{(n-1)!^2}{(2n)!}.$$

To evaluate this series we note that the power series stated in the problem satisfies the differential equation

$$(1 - x^2)y'' - xy' - 2 = 0$$

with the initial conditions $y(0) = y'(0) = 0$ as well as the function $\arcsin^2 x$; hence they coincide with each other. We thus have

$$\zeta(2) = 6\arcsin^2\frac{1}{2} = \frac{\pi^2}{6}.$$

$\square$

SOLUTION 10.7

For any positive integer n it follows from integration by parts that

$$\begin{aligned}
I_n &= \left[\theta\cos^{2n}\theta\right]_{\theta=0}^{\theta=\pi/2} + 2n\int_0^{\pi/2}\theta\sin\theta\cos^{2n-1}\theta\, d\theta \\
&= n\left[\theta^2\sin\theta\cos^{2n-1}\theta\right]_{\theta=0}^{\theta=\pi/2} \\
&\quad - n\int_0^{\pi/2}\theta^2\left(\cos^{2n}\theta - (2n-1)\sin^2\theta\cos^{2n-2}\theta\right)d\theta;
\end{aligned}$$

therefore

$$I_n = n(2n-1)J_{n-1} - 2n^2 J_n.$$

Multiplying this by $2^{2n-1}(n-1)!^2/(2n)!$ we get

$$\frac{\pi}{4n^2} = \frac{4^{n-1}(n-1)!^2}{(2n-2)!}J_{n-1} - \frac{4^n n!^2}{(2n)!}J_n.$$

Adding these formulae from $n = 1$ to $n = m$ we obtain

$$\frac{\pi}{4} \sum_{n=1}^{m} \frac{1}{n^2} = J_0 - \frac{4^m m!^2}{(2m)!} J_m.$$

Since $J_0 = \pi^3/24$, it suffices to show that the second term on the right-hand side of the above expression converges to 0 as $m \to \infty$. To see this, using Jordan's inequality $\pi \sin \theta \geq 2\theta$ valid on the interval $[0, \pi/2]$ we get

$$J_m < \frac{\pi^2}{4} \int_0^{\pi/2} \sin^2 \theta \cos^{2m} \theta \, d\theta = \frac{\pi^2}{4} (I_m - I_{m-1})$$

$$= \frac{\pi^2}{4(m+1)} I_m.$$

Hence

$$0 < \frac{4^m m!^2}{(2m)!} J_m < \frac{4^m m!^2}{(2m)!} \cdot \frac{\pi^2}{4(m+1)} \cdot \frac{(2m)! \pi}{m!^2 2^{2m+1}}$$

$$= \frac{\pi^3}{8(m+1)}$$

and the right-hand side converges to 0 as $m \to \infty$, as required. $\square$

SOLUTION 10.8

Let I be the value of the improper integral stated in the problem. Since both complex numbers $e^{i\theta}$ and $1 + e^{i\theta}$ do not touch the negative real axis as θ varies in the interval $[0, \pi/2)$, we can write

$$\log(2 \cos \theta) = i\theta + \log(1 + e^{-2i\theta})$$

by taking the principal value of the logarithm. Therefore we obtain

$$I = \frac{\pi^2}{8} i + \int_0^{\pi/2} \log(1 + e^{-2i\theta}) \, d\theta.$$

Moreover the Taylor series of $\log(1+z)$ converges on compact sets in $\{|z| = 1, z \neq -1\}$, which is the set with the point $z = -1$ removed from the unit circle centered at the origin. Thus for any sufficiently small $\epsilon > 0$ we have

$$\int_0^{\pi/2-\epsilon} \log(1 + e^{-2i\theta}) \, d\theta = \sum_{n=1}^{\infty} \frac{(-1)^{n-1}}{n} \cdot \frac{1 - (-1)^n e^{2n\epsilon i}}{2ni}.$$

Since the right-hand side now converges absolutely to $-3\zeta(2)i/4$ as $\epsilon \to 0+$, we get

$$I = \frac{i}{8}(\pi^2 - 6\zeta(2)).$$

However I is obviously real and this means that $I = 0$; that is, $\zeta(2) = \pi^2/6$.   □

SOLUTION 10.9

For any x in the interval $(-1, 1)$ we put

$$\phi(x) = \int_{-1}^{1} \frac{dy}{1 + xy} = \frac{1}{x} \log \frac{1 + x}{1 - x}$$

for brevity. Expanding the function $1/(1 + xy)$ into the Taylor series with respect to y, we get

$$\phi(x) = \sum_{n=0}^{\infty} (-1)^n x^n \int_{-1}^{1} y^n \, dy = 2 \sum_{n=0}^{\infty} \frac{x^{2n}}{2n + 1}.$$

Therefore we obtain

$$I_\epsilon = \int_0^{1-\epsilon} \phi(x) \, dx = 2 \sum_{n=0}^{\infty} \frac{(1 - \epsilon)^{2n+1}}{(2n + 1)^2}$$

for any ϵ in $(0, 1)$. The series on the right-hand side clearly converges to $3\zeta(2)/2$ as $\epsilon \to 0+$.

On the other hand, the relation

$$\cos \theta = y + \frac{x}{2}(y^2 - 1)$$

gives a smooth one-to-one correspondence between the interval $[-1, 1]$ in y and the interval $[0, \pi]$ in θ. Hence we obtain

$$\phi(x) = \int_\pi^0 \frac{1}{1 + xy} \cdot \frac{dy}{d\theta} \, d\theta = \int_0^\pi \frac{\sin \theta}{(1 + xy)^2} \, d\theta$$

$$= \int_0^\pi \frac{\sin \theta}{1 + 2x \cos \theta + x^2} \, d\theta.$$

Since

$$\frac{\sin \theta}{1 + 2x \cos \theta + x^2} = \frac{d}{dx} \left(\arctan \frac{x + \cos \theta}{\sin \theta} \right),$$

it follows from this by interchanging the order of integration that

$$I_\epsilon = \int_0^\pi \left(\arctan \frac{1 + \cos\theta - \epsilon}{\sin\theta} - \arctan\cot\theta \right) d\theta$$

$$= \int_0^\pi \arctan \frac{(1-\epsilon)\sin\theta}{1 + (1-\epsilon)\cos\theta} \, d\theta,$$

which converges to

$$\int_0^\pi \arctan \frac{\sin\theta}{1 + \cos\theta} \, d\theta = \int_0^\pi \frac{\theta}{2} \, d\theta = \frac{\pi^2}{4}$$

as $\epsilon \to 0+$, since the integrand is uniformly bounded. □

SOLUTION 10.10

The affine transformation described in the problem clearly rotates the unit square $(0,1) \times (0,1)$ with -45 degrees so that the four vertices $(0,0)$, $(1,0)$, $(1,1)$ and $(0,1)$ are transformed to $(0,0)$, $(1/\sqrt{2}, -1/\sqrt{2})$, $(\sqrt{2},0)$ and $(1/\sqrt{2}, 1/\sqrt{2})$ respectively. Since

$$1 - xy = 1 - \frac{u^2 - v^2}{2}$$

is an even function with respect to v, we have

$$\frac{\zeta(2)}{4} = \int_0^{1/\sqrt{2}} \int_0^u + \int_{1/\sqrt{2}}^{\sqrt{2}} \int_0^{\sqrt{2}-u} \frac{du\,dv}{2 - u^2 + v^2}. \qquad (10.1)$$

Hence we obtain the required formula given in the problem by using

$$\int_0^x \frac{dv}{2 - u^2 + v^2} = \frac{1}{\sqrt{2 - u^2}} \arctan \frac{x}{\sqrt{2 - u^2}}.$$

Substituting u by $\sqrt{2}\sin\theta$ and $\sqrt{2}\cos 2\theta$ in the first and the second integral in (10.1) respectively, we get

$$\frac{\zeta(2)}{4} = \int_0^{\pi/6} \theta \, d\theta + 2 \int_0^{\pi/6} \theta \, d\theta = \frac{\pi^2}{24}.$$

□

SOLUTION 10.11

Let Φ and Δ be the transformation and the triangular region stated in the problem respectively. Obviously Φ maps Δ into the unit open square $(0, 1) \times (0, 1)$ in the xy-plane. Conversely for any x and y in the interval $(0, 1)$ we have the formulae

$$\sin^2\theta = \frac{x^2(1 - y^2)}{1 - x^2 y^2} \quad \text{and} \quad \sin^2\varphi = \frac{y^2(1 - x^2)}{1 - x^2 y^2},$$

from which we can determine θ and φ in $(0, \pi/2)$ respectively. Moreover

$$\cos(\theta + \varphi) = \frac{\sqrt{(1 - x^2)(1 - y^2)}}{1 + xy} > 0$$

implies that $\theta + \varphi < \pi/2$; hence $(\theta, \varphi) \in \Delta$. Since the Jacobian of Φ

$$\begin{vmatrix} \dfrac{\partial x}{\partial \theta} & \dfrac{\partial x}{\partial \varphi} \\ \dfrac{\partial y}{\partial \theta} & \dfrac{\partial y}{\partial \varphi} \end{vmatrix} = \begin{vmatrix} \dfrac{\cos\theta}{\cos\varphi} & \dfrac{\sin\theta}{\cos\varphi}\tan\varphi \\ \dfrac{\sin\varphi}{\cos\theta}\tan\theta & \dfrac{\cos\varphi}{\cos\theta} \end{vmatrix}$$

is equal to $1 - x^2 y^2 > 0$, we conclude that

$$\frac{3}{4}\zeta(2) = \iint_\Delta d\theta d\varphi = \frac{\pi^2}{8}.$$

$\square$

SOLUTION 10.12

Let I be the value of the improper repeated integral in the problem. By integrating first in y we have

$$I = \int_0^\infty \frac{1}{1 + x^2} \left[\arctan xy\right]_{y=0}^{y=1} dx = \int_0^\infty \frac{\arctan x}{1 + x^2} dx,$$

which is equal to

$$\left[\frac{1}{2}\arctan^2 x\right]_{x=0}^{x=\infty} = \frac{\pi^2}{8}.$$

Next by integrating first in x we have

$$I = \frac{1}{2}\int_0^1 \frac{dy}{1 - y^2} \int_0^\infty \left(\frac{2x}{1 + x^2} - \frac{2xy^2}{1 + x^2 y^2}\right) dx$$

$$= \frac{1}{2}\int_0^1 \frac{1}{1 - y^2} \left[\log \frac{1 + x^2}{1 + x^2 y^2}\right]_{x=0}^{x=\infty} dy,$$

which is equal to

$$-\int_0^1 \frac{\log y}{1 - y^2} \, dy.$$

Then by the termwise integration we obtain

$$I = -\sum_{n=0}^{\infty} \int_0^1 y^{2n} \log y \, dy = \sum_{n=0}^{\infty} \frac{1}{(2n+1)^2} = \frac{3}{4}\zeta(2).$$

□

SOLUTION 10.13

Since

$$\left(D\left(-\frac{1}{z}\right) + D(-z) \right)' = \frac{\log(1 + 1/z) - \log(1 + z)}{z} = -\frac{\log z}{z}$$

for any complex $z \in \mathbb{C}\backslash(-\infty, 0]$, we have the functional equation

$$D\left(-\frac{1}{z}\right) + D(-z) = c - \frac{1}{2}\log^2 x$$

for some constant c. To determine the value of c we simply put $z = 1$ in the above formula to obtain

$$c = 2D(-1) = -\zeta(2).$$

On the other hand, putting $z = -1 + \epsilon i$ and taking the limit of $D(-1/z)$ as $\epsilon \to 0+$ we see that $D(-1/z)$, together with $D(-z)$, converges to $D(1) = \zeta(2)$. Hence it follows from the functional equation that

$$2\zeta(2) = -\zeta(2) - \frac{1}{2}(-\pi i)^2;$$

namely $\zeta(2) = \pi^2/6$, as required.

□

Chapter 11

Functions of Several Variables

- Let $f(x, y)$ and $\dfrac{\partial f}{\partial y}(x, y)$ be continuous on $[a, b] \times (c, d)$. Then it follows that

$$\frac{d}{dy} \int_a^b f(x, y)\, dx = \int_a^b \frac{\partial f}{\partial y}(x, y)\, dx$$

for any $y \in (c, d)$. This means that we can interchange the order of differentiation and integration; in other words, we can differentiate under the integral sign.

- If $f(x, y)$ and $\dfrac{\partial f}{\partial y}(x, y)$ are continuous on $[a, \infty) \times (c, d)$,

$$\int_a^\infty f(x, y)\, dx$$

exists, and if

$$\int_a^\infty \frac{\partial f}{\partial y}(x, y)\, dx$$

converges uniformly on compact sets in y, then it holds that

$$\frac{d}{dy} \int_a^\infty f(x, y)\, dx = \int_a^\infty \frac{\partial f}{\partial y}(x, y)\, dx$$

for any y in (c, d).

- For n-tuple of non-negative integers $\boldsymbol{m} = (m_1, m_2, ..., m_n)$ the partial differential operator

$$\frac{\partial^{|\boldsymbol{m}|}}{\partial x_1^{m_1} \partial x_2^{m_2} \cdots \partial x_n^{m_n}}$$

is denoted by $D^{\boldsymbol{m}}$ where $|\boldsymbol{m}| = m_1 + m_2 + \cdots + m_n$ is called the order of $D^{\boldsymbol{m}}$.

- Let f have continuous partial derivatives of order s at each point of an open set U in $\mathbb{R}^n$. If the line segment joining two points $\boldsymbol{x} = (x_1, ..., x_n)$ and $\boldsymbol{a} = (a_1, ..., a_n)$ is contained in U, then there exists a θ in the interval $(0, 1)$ such that

$$f(\boldsymbol{x}) = f(\boldsymbol{a}) + \sum_{k=1}^{s-1} \frac{1}{k!} \left(\sum_{j=1}^{n} (x_j - a_j) \frac{\partial}{\partial x_j} \right)^k f(\boldsymbol{a})$$

$$+ \frac{1}{s!} \left(\sum_{j=1}^{n} (x_j - a_j) \frac{\partial}{\partial x_j} \right)^s f((1 - \theta)\boldsymbol{a} + \theta\boldsymbol{x}),$$

  known as Taylor's formula for functions of several variables.

- For any polynomial $P(x_1, ..., x_n)$ it may sometimes be convenient to express

$$P(\boldsymbol{x}) = \sum_{k_1=0}^{r_1} \cdots \sum_{k_n=0}^{r_n} \frac{1}{\boldsymbol{k}!} D^{\boldsymbol{k}} P(\boldsymbol{a}) (\boldsymbol{x} - \boldsymbol{a})^{\boldsymbol{k}}$$

  where $\boldsymbol{k} = (k_1, ..., k_n)$, $\boldsymbol{k}! = k_1! k_2! \cdots k_n!$,

$$(\boldsymbol{x} - \boldsymbol{a})^{\boldsymbol{k}} = (x_1 - a_1)^{k_1} \cdots (x_n - a_n)^{k_n}$$

  and r_j is the degree of P with respect to x_j.

- Let $u_k(x_1, ..., x_n)$, $1 \le k \le n$ be a smooth transformation from an open region U onto V in $\mathbb{R}^n$. Suppose that the Jacobian

$$J = \begin{vmatrix} \dfrac{\partial u_1}{\partial x_1} & \cdots & \dfrac{\partial u_1}{\partial x_n} \\ \vdots & \ddots & \vdots \\ \dfrac{\partial u_n}{\partial x_1} & \cdots & \dfrac{\partial u_n}{\partial x_n} \end{vmatrix}$$

  does not vanish on U. Then

$$\int_V \cdots \int f(u_1, ..., u_n) \, du_1 \cdots du_n$$

$$= \int_U \cdots \int f(u_1(\boldsymbol{x}), ..., u_n(\boldsymbol{x})) |J| \, dx_1 \cdots dx_n$$

  for any continuous function f on V, if the integral on the left-hand side exists.

After Apéry's discovery (1978) of the irrationality proof of

$$\zeta(3) = \sum_{n=1}^{\infty} \frac{1}{n^3},$$

Beukers (1978) gave another elegant proof using the following improper triple integral:

$$\zeta(3) = \iiint_B \frac{dxdydz}{1 - (1 - xy)z}$$

where B is the unit cube $(0, 1)^3$.

PROBLEM 11.1

Let $\square$ be the hypercube $[0, 1]^n$. For any $f \in C[0, 1]$ show that

$$\lim_{n \to \infty} \int \cdots \int_\square f\left(\frac{x_1 + x_2 + \cdots + x_n}{n}\right) dx_1 dx_2 \cdots dx_n = f\left(\frac{1}{2}\right).$$

This is given in Kac's book (1972).

PROBLEM 11.2

Show that

$$\iint_{0<x<y<\pi} \log|\sin(x - y)|\, dxdy = -\frac{1}{2}\pi^2 \log 2.$$

PROBLEM 11.3

We assign a complex number λ_m to each n-tuple of non-negative integers $m = (m_1, ..., m_n)$ arbitrarily. Show that there exists an $f(x_1, ..., x_n) \in C^\infty(\mathbb{R}^n)$ satisfying

$$D^m f(\mathbf{0}) = \lambda_m$$

for any m, where $\mathbf{0} = (0, ..., 0)$.

The one variable case was shown by Borel (1895). Later Rosenthal (1953) gave a simpler proof by considering

$$g(x) = \sum_{n=0}^{\infty} a_n e^{-|a_n|n!x^2} x^n$$

where a_n is determined according to the given value of $g^{(k)}(0)$. Mirkil (1956) gave a proof for the n-dimensional case.

PROBLEM 11.4

To generalize the method in **SOLUTION 10.11** *consider the 2n-dimensional improper integral:*

$$I_n = \int \cdots \int_\square \frac{dx_1 \cdots dx_{2n}}{1 - (x_1 \cdots x_{2n})^2}$$

where $\square$ *is the open hypercube* $(0, 1)^{2n}$. *Carry out the transformation*

$$x_1 = \frac{\sin \theta_1}{\cos \theta_2}, \quad \ldots, \quad x_{2n-1} = \frac{\sin \theta_{2n-1}}{\cos \theta_{2n}}, \quad x_{2n} = \frac{\sin \theta_{2n}}{\cos \theta_1}$$

to show that

$$\zeta(2n) = \frac{2^{2n-1}}{(2n)!} B_n \pi^{2n}$$

where B_n *is the nth Bernoulli number.*

This is due to Beukers, Calabi and Kolk (1993). See the remark after **PROBLEM 10.11**.

PROBLEM 11.5

Let f be a function of two variables x and y defined on an open region U such that both $\frac{\partial f}{\partial x}(x, y)$ *and* $\frac{\partial f}{\partial y}(x, y)$ *exist on U. If they are totally differentiable at a point* $(a, b) \in U$, *then show that*

$$\frac{\partial^2 f}{\partial x \partial y}(a, b) = \frac{\partial^2 f}{\partial y \partial x}(a, b).$$

This is due to Young (1909) and called the fundamental theorem of differentials. Note that no continuity for the partial derivatives are assumed.

It is easily verified that

$$f(x, y) = \begin{cases} \dfrac{xy(x^2 - y^2)}{x^2 + y^2} & \text{if} \quad (x, y) \neq (0, 0), \\ 0 & \text{if} \quad (x, y) = (0, 0), \end{cases}$$

satisfies

$$\frac{\partial^2 f}{\partial x \partial y}(0, 0) = -1 \quad \text{and} \quad \frac{\partial^2 f}{\partial y \partial x}(0, 0) = 1.$$

Check that this function does not satisfy the conditions stated in the problem.

Solutions for Chapter 11

$\boxed{\text{Solution } 11.1}$

We first compute the average E and the variance V of

$$X = \frac{x_1 + x_2 + \cdots + x_n}{n}.$$

It is easily verified that

$$\begin{cases} E = \int \cdots \int_{B_n} X \, dx_1 dx_2 \cdots dx_n = \frac{1}{2}, \\ V = \int \cdots \int_{B_n} \left(X - \frac{1}{2} \right)^2 dx_1 dx_2 \cdots dx_n = \frac{1}{12n}, \end{cases}$$

where B_n is the n-dimensional unit cube $[0, 1]^n$. Let c_n be an arbitrary positive sequence converging to 0 as $n \to \infty$, and J_n be the subset of B_n satisfying

$$\left| X - \frac{1}{2} \right| \geq c_n.$$

We have immediately

$$\frac{1}{12n} = V \geq c_n^2 \int \cdots \int_{J_n} dx_1 dx_2 \cdots dx_n;$$

hence

$$\int \cdots \int_{J_n} dx_1 dx_2 \cdots dx_n \leq \frac{1}{12n c_n^2}.$$

The right-hand side converges to 0 if, for example, we take $c_n = n^{-1/3}$. Then for any $\epsilon > 0$ we can find a sufficiently large integer N such that

$$\left| f(X) - f\left(\frac{1}{2} \right) \right| < \epsilon$$

on the set $B_n \backslash J_n$ for all integers $n > N$, since

$$\left| X - \frac{1}{2} \right| < n^{-1/3}$$

on the set $B_n \setminus J_n$. Thus,

$$\left| \int \cdots \int_{B_n} f(X)\,dx_1 \cdots dx_n - f\left(\frac{1}{2}\right) \right|$$

$$\leq \int \cdots \int_{J_n} + \int \cdots \int_{B_n \setminus J_n} \left| f(X) - f\left(\frac{1}{2}\right) \right| dx_1 \cdots dx_n$$

$$\leq \frac{M}{6n^{1/3}} + \epsilon,$$

where M is the maximum of $|f(x)|$ on the interval $[0, 1]$. We can take a larger N if necessary such that $M/(6n^{1/3}) < \epsilon$ holds for all $n > N$. $\square$

SOLUTION 11.2

By the affine transformation

$$\begin{pmatrix} x \\ y \end{pmatrix} = \begin{pmatrix} -1/2 & 1/2 \\ 1/2 & 1/2 \end{pmatrix} \begin{pmatrix} u \\ v \end{pmatrix}$$

the triangular region $\{0 < x < y < \pi\}$ is the image of

$$D = \{0 < u < v < 2\pi - u\}$$

with the Jacobian $-1/2$. The double integral in the problem is thus equal to

$$\frac{1}{2} \iint_D \log |\sin u|\, du dv = \int_0^\pi (\pi - u) \log |\sin u|\, du = J.$$

Since $(\pi/2 - u) \log |\sin u|$ is an odd function with respect to $u = \pi/2$, the corresponding integral over the interval $(0, \pi)$ vanishes. Therefore

$$J = \pi \int_0^{\pi/2} \log |\sin u|\, du$$

and we get

$$\frac{2}{\pi} J = \int_0^{\pi/2} \log |\sin u|\, du + \int_0^{\pi/2} \log |\cos u|\, du$$

$$= \int_0^{\pi/2} \log |\sin 2u|\, du - \frac{\pi}{2} \log 2 = \frac{J}{\pi} - \frac{\pi}{2} \log 2,$$

which implies that $J = -\frac{1}{2}\pi^2 \log 2$. $\square$

REMARK. The evaluation of J is due to Euler. Pólya and Szegö (1972) used the so-called Vandermonde determinant

$$
\begin{vmatrix}
1 & 1 & 1 & \cdots & 1 \\
1 & z & z^2 & \cdots & z^{n-1} \\
1 & z^2 & z^4 & \cdots & z^{2(n-1)} \\
\vdots & \vdots & \vdots & \ddots & \vdots \\
1 & z^{n-1} & z^{2(n-1)} & \cdots & z^{(n-1)^2}
\end{vmatrix}
= \prod_{0 \le j < k < n} (z^k - z^j)
$$

to evaluate the value of J. Computing the product of the determinant with $z = \exp(2\pi i/n)$ and its conjugate, they obtained

$$
\prod_{0 \le j < k < n} \left(2 \sin \frac{j-k}{n} \pi \right)^2 = n^n.
$$

The right-hand side comes from computing the product of two matrices directly. Therefore

$$
\frac{\pi^2}{n^2} \sum_{0 \le j < k < n} \log \left| \sin \left(\frac{j\pi}{n} - \frac{k\pi}{n} \right) \right| = \frac{\pi^2}{2n} \log n - \left(1 - \frac{1}{n} \right) \frac{\pi^2}{2} \log 2.
$$

The left-hand side may be the 2-dimensional equally divided Riemann sum for the function $\log|\sin(x-y)|$, although we must verify the convergence of the Riemann sum to the improper double integral.

SOLUTION 11.3

Let $\|x\|$ be the Euclidean distance in $\mathbb{R}^n$ between $x = (x_1, \ldots, x_n)$ and $0 = (0, \ldots, 0)$. Take a C^∞-function $\phi(x)$ defined on $\mathbb{R}^n$ satisfying

$$
\phi(x) = \begin{cases} 1 & \text{for} \quad \|x\| \le 1, \\ 0 & \text{for} \quad \|x\| \ge 2. \end{cases}
$$

For any non-negative integer k we put

$$
f_k(x) = \sum_{|m|=k} \frac{\lambda_m}{m!} x^m \phi(x),
$$

where $m! = m_1! m_2! \cdots m_n!$, $x^m = x_1^{m_1} \cdots x_n^{m_n}$ and the summation runs through $m = (m_1, \ldots, m_n)$ satisfying $|m| = m_1 + m_2 + \cdots + m_n = k$.

Then it can be seen that

$$
D^\ell f_k(x) = \sum_{|m|=k} \frac{\lambda_m}{m!} \sum_{i+j=\ell} \binom{\ell}{i} D^i(x^m) D^j \phi(x)
$$

for any n-tuple of non-negative integers $\boldsymbol{\ell} = (\ell_1, ..., \ell_n)$, where

$$\binom{\boldsymbol{\ell}}{i} = \frac{\ell!}{i!\,(\ell - i)!}.$$

If $m_s \geq i_s$ for all $1 \leq s \leq n$, then

$$D^i(\boldsymbol{x}^m) = \frac{m!}{(m - i)!}\boldsymbol{x}^{m-i}.$$

On the other hand, if $m_s < i_s$ for some s, then clearly $D^i(\boldsymbol{x}^m)$ vanishes. Hence $D^i(\boldsymbol{x}^m)(\boldsymbol{0})$ does not vanish if and only if $i = m$, whence we obtain

$$D^m(\boldsymbol{x}^m)(\boldsymbol{0}) = m!.$$

Since $\phi(\boldsymbol{x})$ is constant on $\|\boldsymbol{x}\| \leq 1$, it is obvious that $D^j\phi(\boldsymbol{0})$ does not vanish if and only if $j = \boldsymbol{0}$. Therefore $D^0\phi(\boldsymbol{0}) = \phi(\boldsymbol{0}) = 1$. We thus have $D^{\boldsymbol{\ell}} f_k(\boldsymbol{0}) = 0$ for any $|\boldsymbol{\ell}| \neq k$ and $D^{\boldsymbol{\ell}} f_k(\boldsymbol{0}) = \lambda_{\boldsymbol{\ell}}$ for any $|\boldsymbol{\ell}| = k$.

Put next

$$M_k = 2^k \max_{|\boldsymbol{\ell}| < k} \max_{\|\boldsymbol{x}\| \leq 2} \left| D^{\boldsymbol{\ell}} f_k(\boldsymbol{x}) \right|$$

and

$$g_k(\boldsymbol{x}) = \frac{1}{M_k^k} f_k\left(M_k x_1, M_k x_2, ..., M_k x_n\right).$$

Since

$$D^{\boldsymbol{\ell}} g_k(\boldsymbol{x}) = M_k^{|\boldsymbol{\ell}|-k} D^{\boldsymbol{\ell}} f_k(M_k \boldsymbol{x}),$$

we have $D^{\boldsymbol{\ell}} g_k(\boldsymbol{0}) = 0$ for any $|\boldsymbol{\ell}| \neq k$ and $D^{\boldsymbol{\ell}} g_k(\boldsymbol{0}) = \lambda_{\boldsymbol{\ell}}$ for any $|\boldsymbol{\ell}| = k$. So $f_k(\boldsymbol{x})$ and $g_k(\boldsymbol{x})$ have the same partial derivatives at the origin. Moreover since

$$\max_{\boldsymbol{x}} \left| D^{\boldsymbol{\ell}} g_k(\boldsymbol{x}) \right| = M_k^{|\boldsymbol{\ell}|-k} \max_{\boldsymbol{x}} \left| D^{\boldsymbol{\ell}} f_k(\boldsymbol{x}) \right| \leq \frac{1}{2^k}$$

for any $|\boldsymbol{\ell}| < k$, this implies that every series obtained by termwise partial differentiation from

$$g(\boldsymbol{x}) = \sum_{k=0}^{\infty} g_k(\boldsymbol{x})$$

converges uniformly in $\mathbb{R}^n$. Therefore $g(\boldsymbol{x}) \in C^{\infty}(\mathbb{R}^n)$ and

$$D^{\boldsymbol{\ell}} g(\boldsymbol{0}) = \sum_{k=0}^{\infty} D^{\boldsymbol{\ell}} g_k(\boldsymbol{0}) = D^{\boldsymbol{\ell}} g_{|\boldsymbol{\ell}|}(\boldsymbol{0}) = \lambda_{\boldsymbol{\ell}}$$

for any n-tuple of non-negative integers $\boldsymbol{\ell}$. $\square$

SOLUTION 11.4

It follows from termwise integration that

$$I_n = \sum_{k=0}^{\infty} \frac{1}{(2k+1)^{2n}} = \left(1 - \frac{1}{2^{2n}}\right)\zeta(2n).$$

Under the transformation described in the problem

$$x_1 = \frac{\sin\theta_1}{\cos\theta_2}, \quad \ldots, \quad x_{2n-1} = \frac{\sin\theta_{2n-1}}{\cos\theta_{2n}}, \quad x_{2n} = \frac{\sin\theta_{2n}}{\cos\theta_1}$$

the hypercube $\square$ is the image of the polytope $\triangle$ defined by

$$\theta_1 + \theta_2 < \frac{\pi}{2}, \quad \ldots, \quad \theta_{2n-1} + \theta_{2n} < \frac{\pi}{2}, \quad \theta_{2n} + \theta_1 < \frac{\pi}{2}.$$

To see this, first observe that $0 < x_1, \ldots, x_{2n} < 1$ for any $(\theta_1, \ldots, \theta_{2n}) \in \triangle$. Conversely, for any $(x_1, \ldots, x_{2n}) \in \square$,

$$\phi(t) = x_{2n}^2\left(1 - x_1^2(1 - \cdots(1 - x_{2n-1}^2(1 - t)\cdots))\right)$$

is a linear function on t and there exists a unique t^* in the open interval $(0, 1)$ satisfying $\phi(t) = t$ since $\phi(0), \phi(1) \in (0, 1)$. Using the point t^* we can determine $\theta_{2n}, \ldots, \theta_1$ in the interval $(0, \pi/2)$ in this order by the formulae

$$\sin\theta_{2n} = \sqrt{t^*}, \quad \sin\theta_{2n-1} = x_{2n-1}\cos\theta_{2n}, \quad \ldots, \quad \sin\theta_1 = x_1\cos\theta_2.$$

The polytope $\triangle$ is thus mapped onto the hypercube $\square$ diffeomorphically, since the Jacobian

$$\begin{vmatrix} u_{1,2} & 0 & 0 & \cdots & 0 & v_{2n,1} \\ v_{1,2} & u_{2,3} & 0 & \cdots & 0 & 0 \\ 0 & v_{2,3} & u_{3,4} & \cdots & 0 & 0 \\ \vdots & \vdots & \vdots & \ddots & \vdots & \vdots \\ 0 & 0 & 0 & \cdots & u_{2n-1,2n} & 0 \\ 0 & 0 & 0 & \cdots & v_{2n-1,2n} & u_{2n,1} \end{vmatrix}$$

where

$$u_{j,k} = \frac{\cos\theta_j}{\cos\theta_k} \quad \text{and} \quad v_{j,k} = \tan\theta_k\frac{\sin\theta_j}{\cos\theta_k},$$

is equal to

$$u_{1,2}u_{2,3}\cdots u_{2n,1} - v_{1,2}v_{2,3}\cdots v_{2n,1} = 1 - (x_1\cdots x_{2n})^2 > 0.$$

Notice that this is equal to the denominator of the integrand considered.

Let $\triangle^*$ be the image of $\triangle$ by the linear contraction in the ratio of $2/\pi$; so,

$$\triangle^* = \{(s_1, ..., s_{2n}) \in \mathbb{R}^{2n}; \ s_1, ..., s_{2n} > 0, \ s_1 + s_2 < 1, ..., s_{2n} + s_1 < 1\}.$$

Thus we have

$$\zeta(2n) = \frac{2^{2n}}{2^{2n} - 1} \int \cdots \int_\triangle d\theta_1 \cdots d\theta_{2n}$$

$$= \frac{\pi^{2n}}{2^{2n} - 1} |\triangle^*|_{2n},$$

where $|X|_m$ denotes the m-dimensional volume of a set X. For each $1 \le j \le 2n$ we put

$$\Omega_j = \{(s_1, ..., s_{2n}) \subset \triangle^*; \ s_j < s_k \ \text{for any} \ k \ne j\}.$$

Obviously Ω_j's are disjoint and congruent mutually so that

$$\overline{\triangle^*} = \bigcup_{j=1}^{2n} \overline{\Omega}_j$$

where $\overline{X}$ denotes the closure of a set X and a fortiori

$$|\triangle^*|_{2n} = 2n |\Omega_{2n}|_{2n}.$$

Now let Σ be the surface of $\overline{\Omega}_{2n}$ defined by $s_{2n} = 0$; namely, Σ is the set of points $(s_1, ..., s_{2n-1}, 0)$ in $\mathbb{R}^{2n}$ satisfying $s_1, ..., s_{2n-1} \ge 0$, $s_1 + s_2 \le 1$, $s_{2n-2} + s_{2n-1} \le 1$ and $s_{2n-1} \le 1$. $\overline{\Omega}_{2n}$ has a pyramid-like shape with the base Σ and the vertex

$$\left(\frac{1}{2}, \frac{1}{2}, ..., \frac{1}{2}\right),$$

since every point in $\overline{\Omega}_{2n}$ can be uniquely expressed as

$$\left((1 - t)s_1 + \frac{t}{2}, ..., (1 - t)s_{2n-1} + \frac{t}{2}, \frac{t}{2}\right)$$

with some $(s_1, ..., s_{2n-1}, 0) \in \Sigma$ and t in the interval $[0, 1]$. This gives the transformation from $\Sigma \times [0, 1]$ onto $\overline{\Omega}_{2n}$ with the Jacobian

$$\begin{vmatrix} 1 - t & 0 & \cdots & 0 & 0 \\ 0 & 1 - t & \cdots & 0 & 0 \\ \vdots & \vdots & \ddots & \vdots & \vdots \\ 0 & 0 & \cdots & 1 - t & 0 \\ \frac{1}{2} - s_1 & \frac{1}{2} - s_2 & \cdots & \frac{1}{2} - s_{2n-1} & \frac{1}{2} \end{vmatrix} = \frac{1}{2}(1 - t)^{2n-1} > 0.$$

Hence we have

$$|\Omega_{2n}|_{2n} = \frac{1}{2} \int \cdots \int_\Sigma \int_0^1 (1-t)^{2n-1} dt \, ds_1 \cdots ds_{2n-1}$$

$$= \frac{1}{4n} \int \cdots \int_\Sigma ds_1 \cdots ds_{2n-1};$$

therefore

$$\zeta(2n) = \frac{\pi^{2n}}{2(2^{2n}-1)} |\Sigma|_{2n-1}.$$

Our problem is thus reduced to the evaluation of the $(2n-1)$-dimensional volume of Σ. It follows from the definition of Σ that

$$|\Sigma|_{2n-1} = \int_0^1 \left(\int_0^{1-s_1} \cdots \left(\int_0^{1-s_{2n-2}} ds_{2n-1} \right) \cdots ds_2 \right) ds_1.$$

We now define a sequence of polynomials $p_m(x)$ inductively by

$$p_{m+1}(x) = \int_0^{1-x} p_m(t) \, dt$$

with the initial condition $p_0(x) = 1$ so that $|\Sigma|_{2n-1} = p_{2n-1}(0)$. $\{p_m(x)\}_{m \geq 0}$ is a bounded sequence on the interval $[0,1]$ and the generating function

$$F(x,\theta) = \sum_{m=0}^\infty p_m(x) \theta^m$$

defined for $|\theta| < 1$ satisfies the second-order differential equation

$$\frac{\partial^2 F}{\partial x^2} + \theta^2 F = 0$$

having the general solution

$$F = a(\theta) \cos \theta x + b(\theta) \sin \theta x.$$

The functions $a(\theta)$ and $b(\theta)$ can be determined by the initial conditions $F(1,\theta) = 1$ and $\partial F / \partial x (0, \theta) = -\theta$; thus,

$$F(x,\theta) = \frac{1 + \sin \theta}{\cos \theta} \cos \theta x - \sin \theta x.$$

Substituting $x = 0$ we obtain

$$\sum_{m=0}^\infty p_m(0) \theta^m = \sec \theta + \tan \theta,$$

Problems and Solutions in Real Analysis

where $\sec\theta$ is the reciprocal of $\cos\theta$. Since $\sec\theta$ is an even function, $p_{2n-1}(0)$ is determined only by the Taylor series of $\tan\theta$ about $\theta = 0$. Indeed,

$$
\tan\theta = -\frac{4i}{e^{4\theta i} - 1} + \frac{2i}{e^{2\theta i} - 1} - i
$$

$$
= -\frac{1}{\theta} + 2i + \sum_{n=1}^{\infty} \frac{2^{4n}}{(2n)!} B_n \theta^{2n-1}
$$

$$
+ \frac{1}{\theta} - i - \sum_{n=1}^{\infty} \frac{2^{2n}}{(2n)!} B_n \theta^{2n-1} - i
$$

$$
= \sum_{n=1}^{\infty} \frac{2^{2n}(2^{2n} - 1)}{(2n)!} B_n \theta^{2n-1},
$$

which implies that

$$
p_{2n-1}(0) = \frac{2^{2n}(2^{2n} - 1)}{(2n)!} B_n.
$$

This completes the proof.   □

REMARK. The method used here to evaluate the volume of the polytope Σ can be found in Macdonald and Nelsen (1979) independently. Also see Elkies (2003).

SOLUTION 11.5

For brevity we write

$$
\Delta_x(x_1, y_1; x_2, y_2) = \frac{\partial f}{\partial x}(x_2, y_2) - \frac{\partial f}{\partial x}(x_1, y_1).
$$

Since $\dfrac{\partial f}{\partial x}$ is totally differentiable at the point (a, b), we have

$$
\Delta_x(a, b; a + \epsilon, b + \eta) = \epsilon \frac{\partial^2 f}{\partial x^2}(a, b) + \eta \frac{\partial^2 f}{\partial y \partial x}(a, b) + \delta(\epsilon, \eta) \tag{11.1}
$$

where $\delta(\epsilon, \eta)/\sqrt{\epsilon^2 + \eta^2}$ converges to 0 as ϵ and η tend to 0 in any manner whatever.

On the other hand, the left-hand side of (11.1) is equal to

$$
\Delta_x(a + \epsilon, b; a + \epsilon, b + \eta) + \Delta_x(a, b; a + \epsilon, b)
$$

$$
= \Delta_x(a + \epsilon, b; a + \epsilon, b + \eta) + \epsilon \frac{\partial^2 f}{\partial x^2}(a, b) + \delta_1(\epsilon), \tag{11.2}
$$

where $\delta_1(\epsilon)/\epsilon$ converges to 0 as $\epsilon \to 0$. Combining (11.1) and (11.2) we get

$$\Delta_x(a + \epsilon, b; a + \epsilon, b + \eta) = \eta \frac{\partial^2 f}{\partial y \partial x}(a, b) + \delta_2(\epsilon, \eta).$$

Similarly one can show, by considering $\dfrac{\partial f}{\partial y}$ and interchanging x and y, that

$$\Delta_y(a, b + \eta; a + \epsilon, b + \eta) = \epsilon \frac{\partial^2 f}{\partial x \partial y}(a, b) + \delta_3(\epsilon, \eta)$$

where $\delta_2(\epsilon, \eta)$ and $\delta_3(\epsilon, \eta)$ have the same property as $\delta(\epsilon, \eta)$.

Since the function $\phi(y) = f(a + \epsilon, y) - f(a, y)$ is differentiable in the interval $(b, b + \eta)$, it follows from the mean value theorem that the double increment

$$\Delta = f(a + \epsilon, b + \eta) - f(a, b + \eta) - f(a + \epsilon, b) + f(a, b)$$

is equal to

$$\phi(b + \eta) - \phi(b) = \eta \phi'(b + \theta \eta)$$

for some $\theta \in (0, 1)$. This can be written as $\eta \Delta_y(a, b+\theta\eta; a+\epsilon, b+\theta\eta)$ and therefore

$$\Delta = \epsilon \eta \frac{\partial^2 f}{\partial x \partial y}(a, b) + \eta \delta_3(\epsilon, \theta\eta).$$

Note that θ depends on ϵ and η. Similarly we obtain

$$\Delta = \epsilon \eta \frac{\partial^2 f}{\partial y \partial x}(a, b) + \epsilon \delta_2(\theta'\epsilon, \eta)$$

for some $\theta' \in (0, 1)$. Thus we have

$$\frac{\partial^2 f}{\partial x \partial y}(a, b) = \frac{\partial^2 f}{\partial y \partial x}(a, b)$$

by taking $\epsilon = \eta$ and tending ϵ to 0. $\qquad\qquad\qquad\square$

.

Chapter 12

Uniform Distribution

Given a sequence $\{a_n\}_{n\geq 1}$ contained in the unit interval $I = [0, 1]$, let $N_n(J)$ be the number of $k \in [1, n]$ satisfying $a_k \in J$, and $|J|$ be the length of J.

- $\{a_n\}_{n\geq 1}$ is said to be uniformly distributed, or equidistributed, on I provided that, for any subinterval J

$$\mu_n(J) = \frac{N_n(J)}{n}$$

converges to $|J|$ as $n \to \infty$. In other words, the probability of finding a_n in J is proportional to $|J|$.

- Obviously we have $\mu_n(I) = 1$ and

$$\mu_n(J \cup J') = \mu_n(J) + \mu_n(J')$$

for any disjoint intervals J and J' in I.

H. Weyl (1885–1955) introduced the notion of uniform distribution in a general form in Weyl (1916).

PROBLEM 12.1

For any positive integer m and any interval $J \subset [0, 1]$ satisfying $|J| < 1/m$ suppose that

$$\limsup_{n \to \infty} \mu_n(J) \leq \frac{1}{m}.$$

Show that the sequence $\{a_n\}_{n\geq 1}$ is uniformly distributed in $[0, 1]$.

172   *Problems and Solutions in Real Analysis*

PROBLEM 12.2

For any irrational number α show that the fractional parts of αn, which is denoted by $\{\alpha n\}$, are uniformly distributed in the interval $[0, 1]$.

The above two are due to Callahan (1964). His argument is very simple because he makes use neither of continued fractions or of exponential sums. Here we present one more simple application of **PROBLEM 12.1**.

PROBLEM 12.3

Suppose that $f \in C^1(0, \infty)$ satisfies $f(x) > 0$, $f'(x) > 0$ for $x > 0$ and $f(x)/x$ diverges to ∞ as $x \to \infty$. Suppose further that

$$\lim_{k \to \infty} \frac{M_k}{m_k} = 1$$

where M_k and m_k are the maximum and the minimum of $f'(x)$ on the interval $[k, k+1]$ respectively. Let f^{-1} be the inverse function of f. Show then that the fractional parts of $f^{-1}(n)$ are uniformly distributed in the interval $[0, 1]$.

For example, the fractional parts of n^α and $\log^\beta n$ are uniformly distributed on $[0, 1]$ when $0 < \alpha < 1$ and $\beta > 1$ respectively.

If f^{-1} grows as fast as or faster than polynomials, the problem becomes much more difficult. For example, it is not known whether the fractional parts of e^n or $(3/2)^n$ are uniformly distributed. It is worth while noting that the investigation of an upper bound for the fractional part of $(3/2)^n$ is closely related to that of the form of $g(k)$ in Waring problem.

However we know that if θ is the Pisot number, then the distance of θ^n from $\mathbb{Z}$ converges to 0 as $n \to \infty$. The Pisot numbers were studied by Pisot (1938, 1946) and by Vijayaraghavan (1940–42, 1948) independently and are sometimes called the PV numbers.

PROBLEM 12.4

A sequence $\{a_n\}_{n \geq 1}$ contained in the interval $[0, 1]$ is uniformly distributed if and only if

$$\lim_{n \to \infty} \frac{1}{n} \sum_{k=1}^{n} f(a_k) = \int_0^1 f(x)\,dx \qquad (12.1)$$

holds for any continuous function $f(x)$ defined on $[0, 1]$.

PROBLEM 12.5

Show that the fractional parts of a given sequence $\{a_n\}_{n \geq 1}$ *is uniformly distributed if and only if*

$$\lim_{n \to \infty} \frac{1}{n} \sum_{k=1}^{n} \exp(2\pi i m a_k) = 0 \qquad (12.2)$$

holds for every positive integer m.

This is due to Weyl (1916) and known as the Weyl criterion.

PROBLEM 12.6

Let α *be an arbitrary positive irrational number. Show that the unit circle* $|z| = 1$ *is the natural boundary of the power series*

$$f(z) = \sum_{n=1}^{\infty} [\alpha n] z^n.$$

This is due to Hecke (1921). The function $f(z)$ is sometimes called the Hecke-Mahler function, being related to Mahler's function of two variables:

$$F_\alpha(w, z) = \sum_{n=1}^{\infty} \sum_{m=1}^{[\alpha n]} w^m z^n$$

by the relation $F_\alpha(1, z) = f(z)$. Various arithmetical properties of the values of $f(z)$ can be studied by virtue of certain functional equations satisfied by $F_\alpha(w, z)$, $F_{1/\alpha}(z, w)$ and $F_{k+\alpha}(w, z)$.

The author (1982) encountered this function in the study of a mathematical neuron model as a special case of Caianiello's equation (1961).

Solutions for Chapter 12

SOLUTION 12.1

Let ϵ be an arbitrary positive number. For any interval J in I with $|J| < 1$ take a rational number $p/q < 1$ satisfying $|J| < p/q < |J| + \epsilon$. We divide J into p equal parts and name them $J_1, J_2, \ldots, J_p$ from left to right. Since $|J_k| < 1/q$ we have

$$\limsup_{n \to \infty} \mu_n(J) \le \sum_{k=1}^{p} \limsup_{n \to \infty} \mu_n(J_k)$$
$$\le \frac{p}{q} < |J| + \epsilon;$$

therefore $\limsup_{n \to \infty} \mu_n(J) \le |J|$ since ϵ is arbitrary.

On the other hand, the set $I \setminus J$ is either an interval or a union of two disjoint intervals; so let K_0 and K_1 be such intervals (the latter may be empty). Then

$$\liminf_{n \to \infty} \mu_n(J) = 1 - \limsup_{n \to \infty} \mu_n(K_0 \cup K_1)$$
$$\ge 1 - \limsup_{n \to \infty} \mu_n(K_0) - \limsup_{n \to \infty} \mu_n(K_1)$$
$$\ge 1 - |K_0| - |K_1| = |J|,$$

which implies that $\lim_{n \to \infty} \mu_n(J) = |J|$, as required.   □

SOLUTION 12.2

Let J be any interval in $[0, 1]$ with $|J| < 1/m$ for some integer $m \ge 2$. Since $\{\alpha n\}$ forms a dense set in $[0, 1]$ (this follows easily from the pigeon hole principle), there exists a sufficiently large integer k satisfying

$$|J| < \{\alpha k\} < \frac{1 - |J|}{m - 1}.$$

Put now $J_0 = J$ and let J_1 be the shifted interval of J_0 by $\{\alpha k\}$. We continue this process up to J_{m-1}; that is,

$$J_j \equiv J + j\{\alpha k\} \pmod 1$$

for $0 \leq j < m$. Since the distance between the left end point of J_0 and the right end point of J_{m-1} measured in the positive direction is

$$m|J| + (m-1)(\{\alpha k\} - |J|) < |J| + (m-1)\frac{1 - |J|}{m-1} = 1,$$

the intervals $J_0, \ldots, J_{m-1}$ are mutually disjoint; therefore

$$\mu_n(J_0) + \mu_n(J_1) + \cdots + \mu_n(J_{m-1}) \leq 1.$$

Moreover, since $j\{\alpha k\} \equiv \{\alpha jk\}$ (mod 1), it can be easily seen that $\{\alpha \ell\} \in J$ if and only if $\{\alpha(\ell + jk)\} \in J_j$ for any $0 \leq j < m$. We thus have

$$|N_n(J) - N_n(J_j)| \leq 2jk,$$

which implies that

$$1 \geq \sum_{j=0}^{m-1} \mu_n(J_j) \geq \sum_{j=0}^{m-1}\left(\mu_n(J) - \frac{2jk}{n}\right)$$

$$= m\mu_n(J) - \frac{km(m-1)}{n}.$$

Since m and k are independent of n, the superior limit of $\mu_n(J)$ as $n \to \infty$ is less than or equal to $1/m$. $\qquad\square$

$\boxed{\text{SOLUTION 12.3}}$

We first show that

$$\lim_{x \to \infty} \frac{f(x+1)}{f(x)} = 1.$$

To see this, by virtue of l'Hôpital's rule, it suffices to show

$$\lim_{x \to \infty} \frac{f'(x+1)}{f'(x)} = 1.$$

For any $\epsilon > 0$ take a large integer k_0 satisfying $M_k/m_k < 1 + \epsilon$ for all $k \geq k_0$. Then for any $x \geq k_0$,

$$\frac{f'(x+1)}{f'(x)} \leq \frac{M_{[x]+1}}{m_{[x]}} < (1 + \epsilon)\frac{m_{[x]+1}}{m_{[x]}} \leq (1 + \epsilon)\frac{M_{[x]}}{m_{[x]}} < (1 + \epsilon)^2.$$

Similarly we can see that the left-hand side is larger than $(1 + \epsilon)^{-2}$. This implies the limit exists and is equal to 1 since ϵ is arbitrary.

Now, for any interval $J = [a, b)$ contained in the unit interval $[0, 1]$, we give an upper estimate for $N_n(J)$. For any $\epsilon > 0$ we take a large integer k_1 satisfying $k < \epsilon f(k)$,

$$\frac{M_k}{m_k} < 1 + \epsilon \quad \text{and} \quad \frac{f(k + 1)}{f(k)} < 1 + \epsilon$$

for all $k \geq k_1$. For any integer n greater than $f(k_1 + a)$ we take a unique integer $K_n \geq k_1$ satisfying

$$f(K_n + a) \leq n < f(K_n + a + 1).$$

Let $v(k)$ be the number of ℓ's satisfying

$$f(k + a) \leq \ell < f(k + b),$$

which is equivalent to $f^{-1}(\ell) - k \in J$; therefore

$$N_n(J) \leq f(k_1 + a) + \sum_{k=k_1}^{K_n} v(k).$$

On the other hand, it follows from the mean value theorem that

$$\begin{aligned} v(k) &\leq [f(b + k)] - [f(a + k)] + 1 \\ &\leq f(b + k) - f(a + k) + 2 \\ &= (b - a)f'(\xi_k) + 2 \end{aligned}$$

for some ξ_k in the interval $(a + k, b + k)$. Hence

$$\begin{aligned} v(k) &\leq (b - a)M_k + 2 \\ &< (1 + \epsilon)(b - a)m_k + 2 \\ &\leq (1 + \epsilon)(b - a)(f(k + 1) - f(k)) + 2; \end{aligned}$$

thus

$$N_n(J) \leq 2K_n + c(k_1, f) + (1 + \epsilon)(b - a)f(K_n + 1).$$

where $c(k_1, f)$ is a constant depending only on k_1 and f. Using $K_n < \epsilon f(K_n) \leq \epsilon n$ and $f(K_n + 1) < (1 + \epsilon)f(K_n) \leq (1 + \epsilon)n$, which follow from the choice of K_n, we get

$$\mu_n(J) < 2\epsilon + (1 + \epsilon)^2(b - a) + O\left(\frac{1}{n}\right),$$

which implies that the superior limit of $\mu_n(J)$ is less than or equal to $|J|$. $\qquad \square$

SOLUTION 12. 4

For any interval J contained in $I = [0, 1]$ the discontinuous function of the first kind

$$\chi_J(x) = \begin{cases} 1 & \text{for} \quad x \in J, \\ 0 & \text{for} \quad x \notin J, \end{cases}$$

is called the characteristic function of J. It then follows from the definition that $\{a_n\}_{n \geq 1}$ is uniformly distributed in I if and only if

$$\lim_{n \to \infty} \frac{1}{n} \sum_{k=1}^{n} \chi_J(a_k) = \int_0^1 \chi_J(x)\, dx \tag{12.3}$$

holds for any subinterval J contained in I.

Since every step function is a finite combination of characteristic functions, Formula (12. 3) holds also for any step functions. It is now clear from the definition of uniform continuity that for any $f \in C(I)$ and any $\epsilon > 0$ there exists a step function $\phi(x)$ satisfying

$$\sup_{x \in I} |f(x) - \phi(x)| < \epsilon.$$

Thus we obtain

$$\left| \frac{1}{n} \sum_{k=1}^{n} f(a_k) - \int_0^1 f(x)\, dx \right| \leq \left| \frac{1}{n} \sum_{k=1}^{n} \phi(a_k) - \int_0^1 \phi(x)\, dx \right| + 2\epsilon.$$

Since the superior limit, as $n \to \infty$, of the left-hand side is less than or equal to 2ϵ, (12. 1) holds for any $f \in C(I)$.

Conversely suppose that (12. 1) holds for any $f \in C(I)$. For any subinterval J contained in I and any $\epsilon > 0$ consider two piecewise linear trapezoidal functions $f_-(x)$ and $f_+(x)$ satisfying

$$f_-(x) \leq \chi_J(x) \leq f_+(x)$$

for any $x \in I$ and $f_\pm(x)$ differ from $\chi_J(x)$ only in some ϵ-neighborhoods of two end points of J. Then obviously

$$\frac{1}{n} \sum_{k=1}^{n} f_-(a_k) \leq \frac{1}{n} \sum_{k=1}^{n} \chi_J(a_k) \leq \frac{1}{n} \sum_{k=1}^{n} f_+(a_k)$$

and the both sides converge to $\int_0^1 f_\pm(x)\,dx$ as $n \to \infty$ respectively. Since

$$\left| \int_0^1 f_\pm(x)\,dx - \int_0^1 \chi_J(x)\,dx \right| \le 2\epsilon,$$

we can conclude that (12.3) holds for any subinterval J in I. ☐

SOLUTION 12.5

By virtue of the previous problem it suffices to show the sufficiency. Suppose that (12.2) holds for any positive integer m. Then clearly we have

$$\lim_{n\to\infty} \frac{1}{n} \sum_{k=1}^{n} T(a_k) = d_0 = \int_0^1 T(x)\,dx$$

for any trigonometric polynomial

$$T(x) = \sum_{j=1}^{p} c_j \sin(2\pi j x) + \sum_{j=0}^{q} d_j \cos(2\pi j x).$$

In the previous proof one can assume in addition that $f_\pm(0) = f_\pm(1)$ respectively. Therefore such $f_\pm$ can be approximated uniformly by some trigonometric polynomials and in consequence (12.3) holds for any subinterval J. This completes the proof. ☐

SOLUTION 12.6

We will show that

$$z_m = e^{2\pi i \alpha m} = e^{2\pi i \{\alpha m\}}$$

on the unit circle $|z| = 1$ is a singular point of $f(z)$ for any positive integer m. As is seen in **PROBLEM 12.2** such points form a dense subset on the unit circle. Putting $a_n = \{\alpha n\} e^{2\pi i \alpha m n}$ and $\sigma_n = a_1 + a_2 + \cdots + a_n$ we have

$$f(r z_m) = \sum_{n=1}^{\infty} \{\alpha n\} r^n e^{2\pi i \alpha m n} = \sum_{n=1}^{\infty} a_n r^n$$

for $0 < r < 1$ and

$$\sigma_n = \sum_{k=1}^{n} \{\alpha k\} r^n e^{2\pi i m \{\alpha k\}} = \sum_{k=1}^{n} \phi(\{\alpha k\}),$$

where $\phi(x) = xe^{2\pi imx}$ is a continuous function on the interval $[0, 1]$. By virtue of
PROBLEM 12.2 and of **PROBLEM 12.4** the sequence σ_n/n converges to

$$\int_0^1 \phi(x)\, dx = \frac{1}{2\pi im}$$

as $n \to \infty$. Let τ_n be the remainder term; that is,

$$\sigma_n = \frac{n}{2\pi im} + \tau_n.$$

For any $\epsilon > 0$ we take a sufficiently large integer n_0 satisfying $|\tau_n| < \epsilon n$ for any
$n \geq n_0$. Hence

$$\frac{f(rz_m)}{1-r} = \sum_{n=1}^{\infty} a_n r^n \sum_{k=0}^{\infty} r^k = \sum_{n=1}^{\infty} \sigma_n r^n$$

$$= \sum_{n=1}^{\infty} \left(\frac{n}{2\pi im} + \tau_n \right) r^n$$

$$= \frac{1}{2\pi im} \cdot \frac{r}{(1-r)^2} + \sum_{n=1}^{\infty} \tau_n r^n.$$

Multiplying the both sides by $(1-r)^2$ we get

$$\left| (1-r) f(rz_m) - \frac{1}{2\pi im} \right| \leq \frac{1-r}{2\pi m} + (1-r)^2 \sum_{n=1}^{\infty} |\tau_n| r^n$$

$$< \frac{1-r}{2\pi m} + (1-r)^2 \sum_{n<n_0} |\tau_n|$$

$$+ \epsilon (1-r)^2 \sum_{n \geq n_0} n r^n.$$

Since the third term on the right-hand side is clearly less than ϵ, the radial limit
of $(z - z_m) f(z)$, as $z \to z_m$ within the unit disk, is equal to $z_m/(2\pi im) \neq 0$. This
implies that z_m is a singular point of $f(z)$, as required. $\square$

Chapter 13

Rademacher Functions

Any real number x in the interval $[0, 1)$ can be expanded into a series of

$$x = \frac{s_1(x)}{2} + \frac{s_2(x)}{2^2} + \cdots + \frac{s_n(x)}{2^n} + \cdots,$$

known as the dyadic or binary expansions of x. Each $s_n(x)$ is a discontinuous function of the first kind taking the value 0 or 1 only. To ensure uniqueness we adopt that the expansion in which all digits are 0 after some place for any irreducible fraction whose denominator is a power of 2. Note that each function $s_n(x)$ is left-continuous.

- The functions defined by

$$r_n(x) = 1 - 2s_n(x)$$

are called Rademacher functions, which form an orthogonal system on the interval $[0, 1]$. This system is not complete but very useful as a sample space of the probability events of coin tossing.

Rademacher functions were studied in Rademacher (1922). According to Grosswald (1980), Rademacher wrote a sequel containing the completion of the system, but he decided not to publish taking Schur's opinion. Soon after Walsh (1923) published some closely relevant results which he obtained independently. Walsh's complete system of orthogonal functions $\{w_n(x)\}$ is defined as

$$w_n(x) = r_{v_1+1}(x) r_{v_2+1}(x) \cdots r_{v_k+1}(x)$$

where $n = 2^{v_1} + 2^{v_2} + \cdots + 2^{v_k}$ with integers $0 \le v_1 < v_2 < \cdots < v_k$.

- We have

$$1 - 2x = \sum_{n=1}^{\infty} \frac{r_n(x)}{2^n}.$$

- If we define

$$r(x) = \begin{cases} 1 & \text{for} \quad 0 \le x < 1, \\ -1 & \text{for} \quad 1 \le x < 2, \end{cases}$$

and extend this to $\mathbb{R}$ as a periodic function with period 2, then the Rademacher function can be written as $r_n(x) = r(2^n x)$, or since $r(x) = (-1)^{[x]}$, we can write

$$r_n(x) = (-1)^{[2^n x]}.$$

The Rademacher function is also defined by

$$\text{sgn}(\sin 2^n \pi x)$$

in some books, which differs from ours only at a set of countable points; so no influence on integrals.

- We use the abbreviation

$$I_n[f(x)] = \int_0^1 f\left(\sum_{k=1}^{n} r_k(x)\right) dx$$

for any positive integer n and any function $f(x)$ defined on $\mathbb{Z}$.

Many problems in this chapter are given in Kac's fruitful monograph (1972).

PROBLEM 13.1

For any real numbers c_1, c_2, ..., c_n show that

$$\int_0^1 \cos\left(\sum_{k=1}^{n} c_k r_k(x)\right) dx = \prod_{k=1}^{n} \cos c_k.$$

Rademacher Functions   183

PROBLEM 13.2

For any positive integers $k_1 < k_2 < \cdots < k_n$ show that

$$\int_0^1 r_{k_1}(x)\, r_{k_2}(x) \cdots r_{k_n}(x)\, dx = 0.$$

If $k_1, k_2, ..., k_n$ are arbitrary positive integers, then the corresponding integral is equal to 1 if and only if the number of k_j satisfying $k_\ell = k_j$ is even for all $1 \le \ell \le n$; otherwise the integral vanishes. For example, we have for any $j \ne k$,

$$\int_0^1 (r_j(x) + r_k(x))^{2m}\, dx = \binom{2m}{0} + \binom{2m}{2} + \cdots + \binom{2m}{2m}$$

$$= \frac{(1+x)^{2m} + (1-x)^{2m}}{2}\Bigg|_{x=1}$$

$$= 2^{2m-1}.$$

PROBLEM 13.3

Compute $I_n[x^2]$ and $I_n[x^4]$.

PROBLEM 13.4

For any non-negative integer m show that $I_n[x^{2m}]$ is a polynomial in n of degree m with integer coefficients whose leading coefficient is

$$\frac{(2m)!}{2^m m!} = 1 \cdot 3 \cdot 5 \cdots (2m-1).$$

PROBLEM 13.5

Let s be an arbitrary real number. Prove that

$$I_n[e^{s|x|}] < 2\left(\frac{e^s + e^{-s}}{2}\right)^n.$$

PROBLEM 13.6

Show that

$$\lim_{n\to\infty} \frac{I_n[|x|]}{\sqrt{n}} = \sqrt{2/\pi}.$$

Hint: Use

$$|s| = \frac{2}{\pi} \int_0^\infty \frac{1-\cos(sx)}{x^2}\,dx$$

to obtain

$$I_n[|x|] = \frac{2}{\pi} \int_0^\infty \frac{1-\cos^n x}{x^2}\,dx.$$

The readers may observe that the limit of

$$\frac{I_n[|x|^{2\lambda}]}{n^\lambda},$$

as $n \to \infty$, exists when λ is a positive integer or $1/2$. Any other example?

PROBLEM 13.7

For any $\epsilon > 0$ show that the series

$$\sum_{n=1}^\infty \frac{1}{n^{2+\epsilon}} \exp\left(\sqrt{\frac{2\log n}{n}}\left|\sum_{k=1}^n r_k(x)\right|\right)$$

converges almost everywhere.

Rademacher Functions 185

Solutions for Chapter 13

SOLUTION 13.1

The left-hand side of the given equality, say I, can be written as

$$I = \Re \int_0^1 \exp\left(i \sum_{k=1}^n c_k r(2^k x)\right) dx$$

$$= \frac{1}{2} \Re \int_0^2 e^{i c_1 r(t)} \phi(t)\, dt,$$

where

$$\phi(t) = \exp\left(i \sum_{k=2}^n c_k r(2^{k-1} t)\right) = \exp\left(i \sum_{k=1}^{n-1} c_{k+1} r_k(t)\right)$$

is clearly a function with period 1. Therefore

$$I = \Re \int_0^1 \frac{e^{i c_1} + e^{-i c_1}}{2} \phi(t)\, dt$$

$$= (\cos c_1) \times \Re \int_0^1 \phi(t)\, dt.$$

By repeating this process we may get the desired equality. □

SOLUTION 13.2

The left-hand side of the given equality, say J, can be written as

$$J = \int_0^1 r(2^{k_1} x) r(2^{k_2} x) \cdots r(2^{k_n} x)\, dx$$

$$= \frac{1}{2^{k_1}} \int_0^{2^{k_1}} r(t) \phi(t)\, dt,$$

where $\phi(t) = r(2^{k_2 - k_1} t) \cdots r(2^{k_n - k_1} t)$ is a function with period 1; hence

$$2^{k_1} J = \sum_{j=0}^{2^{k_1 - 1} - 1} \int_{2j}^{2j+2} r(t) \phi(t)\, dt = 0.$$

□

SOLUTION 13.3

The integral of $r_i(x)r_j(x)$ over the interval $[0, 1]$ vanishes if $i \neq j$ and is equal to 1 if $i = j$. Expanding

$$\left(r_1(x) + r_2(x) + \cdots + r_n(x)\right)^2$$

we get immediately $I_n[x^2] = n$. Similarly the integral of

$$r_i(x)r_j(x)r_k(x)r_\ell(x)$$

over $[0, 1]$ is equal to 1 if and only if either $i = j = k = \ell$ (n kinds of 'four cards') or $\binom{n}{2}$ kinds of 'two pairs'; otherwise the integral vanishes. Therefore

$$I_n[x^4] = n + \binom{4}{2}\binom{n}{2}.$$

$\square$

REMARK. From the expression of $I_n[x^4]$ we can conclude that

$$\phi_n(x) = \frac{1}{n} \sum_{k=1}^{n} r_k(x)$$

converges to 0 almost everywhere as $n \to \infty$, since

$$\sum_{n=1}^{\infty} \int_0^1 \phi_n^4(x)\,dx < \infty$$

and so $\sum \phi_n^4(x)$ has a finite sum almost everywhere by the monotone convergence theorem of Beppo Levi.

SOLUTION 13.4

It is clear that

$$r_1(x) + r_2(x) + \cdots + r_n(x) = n - 2k$$

where k is the number of r_j satisfying $r_j(x) = -1$. Then the number of subintervals of length 2^{-n} on which $r_j(x) = -1$ is equal to the number of ways of picking k unordered -1's from n possibilities. Hence we obtain in general

$$I_n[f(x)] = \frac{1}{2^n} \sum_{k=0}^{n} \binom{n}{k} f(n - 2k);$$

so we put

$$A_m(n) = I_n[x^{2m}] = \frac{1}{2^n} \sum_{k=0}^{n} \binom{n}{k} (n - 2k)^{2m}.$$

We now introduce the rational function

$$R_n(x) = \left(x + \frac{1}{x} \right)^n = \sum_{k=0}^{n} \binom{n}{k} x^{n-2k}$$

and the differential operator $\delta = x\dfrac{d}{dx}$ so that

$$A_m(n) = \frac{1}{2^n} \delta^{2m}(R_n) \Big|_{x=1}.$$

Since $\delta^2(R_n) = n^2 R_n - 4n(n - 1)R_{n-2}$, we have

$$A_{m+1}(n) = \frac{1}{2^n} \delta^{2m}(n^2 R_n - 4n(n - 1)R_{n-2}) \Big|_{x=1}$$
$$= n^2 A_m(n) - n(n - 1)A_m(n - 2)$$

for any integers $m \geq 0$ and $n \geq 3$. This recursion formula enables us to determine all $A_m(n)$ from the initial conditions $A_0(n) = 1$, $A_m(1) = 1$ and $A_m(2) = 2^{2m-1}$. This also gives a solution to **PROBLEM 13.3** but the proof given there is much simpler.

By the recursion formula it is easily seen that $A_m(n)$ is a polynomial in n of degree m with integer coefficients whose leading coefficient is

$$1 \cdot 3 \cdot 5 \cdots (2m - 1).$$

Moreover every polynomial $A_m(n)$ has a factor n if $m \geq 1$. $\square$

$\boxed{\text{SOLUTION 13.5}}$

Obviously we have

$$I_n[e^{s|x|}] < I_n[e^{sx}] + I_n[e^{-sx}] = J_n.$$

Since

$$g_n(x) = \exp\left(s \sum_{k=1}^{n} r_k(x) \right)$$

is a periodic function with period 1, we get

$$J_n = \frac{1}{2}\left(\int_0^2 e^{sr(t)}g_{n-1}(t)\,dt + \int_0^2 \frac{e^{-sr(t)}}{g_{n-1}(t)}\,dt\right)$$

$$= \frac{e^s + e^{-s}}{2}\left(I_{n-1}[e^{sx}] + I_{n-1}[e^{-sx}]\right);$$

therefore

$$\frac{J_n}{J_{n-1}} = \frac{e^s + e^{-s}}{2}.$$

Solving this with $J_0 = 2$ we may get the desired estimate.   □

SOLUTION 13.6

To show the integral formula representing $|s|$ we can assume $s > 0$. By the substitution $t = sx$ we get

$$\int_0^\infty \frac{1 - \cos(sx)}{x^2}\,dx = s\int_0^\infty \frac{1 - \cos t}{t^2}\,dt$$

$$= s\int_0^\infty \frac{\sin t}{t}\,dt = \frac{\pi}{2}s,$$

where the evaluation of the last improper integral was given in SOLUTION 7.11. Interchanging the order of integrations and using PROBLEM 13.1 we have

$$I_n[|x|] = \frac{2}{\pi}\int_0^1 dx \int_0^\infty \frac{1 - \cos(s(r_1(x) + r_2(x) + \cdots + r_n(x)))}{s^2}\,ds$$

$$= \frac{2}{\pi}\int_0^\infty \frac{1 - I_n[\cos(sx)]}{s^2}\,ds$$

$$= \frac{2}{\pi}\int_0^\infty \frac{1 - \cos^n s}{s^2}\,ds,$$

as stated.

For a real parameter $0 \le \epsilon < 1$ we consider the function

$$\varphi_\epsilon(x) = \frac{x^2}{2(1 - \epsilon)} + \log \cos x$$

for $0 \le x < \pi/2$. Clearly $\varphi_\epsilon(0) = \varphi'_\epsilon(0) = 0$ and

$$\varphi''_\epsilon(x) = \frac{1}{1 - \epsilon} - \frac{1}{\cos^2 x}.$$

Hence it follows that $\varphi_0(x) < 0 < \varphi_\epsilon(x)$ on the interval $(0, \alpha(\epsilon))$ if $\epsilon > 0$ where

$$\alpha(\epsilon) = \arccos\sqrt{1 - \epsilon}\,;$$

therefore

$$\exp\left(-\frac{x^2}{2(1-\epsilon)}\right) < \cos x < \exp\left(-\frac{x^2}{2}\right)$$

for $0 < x < \alpha(\epsilon)$. We then have

$$K(2 - 2\epsilon) + \beta(\epsilon) < \frac{\pi}{2} I_n[|x|] < K(2) + \beta(\epsilon) \tag{13.1}$$

where

$$K(\sigma) = \int_0^{\alpha(\epsilon)} \frac{1 - \exp(-ns^2/\sigma)}{s^2}\, ds \quad \text{and} \quad \beta(\epsilon) = \int_{\alpha(\epsilon)}^{\infty} \frac{1 - \cos^n s}{s^2}\, ds.$$

Under the substitution $t = \sqrt{n/\sigma}\, s$ we get

$$K(\sigma) = \int_0^{\alpha(\epsilon)} \frac{1 - \exp(-ns^2/\sigma)}{s^2}\, ds$$

$$= \sqrt{n/\sigma} \int_0^{\tau(\epsilon)} \frac{1 - e^{-t^2}}{t^2}\, dt,$$

which is asymptotic to

$$\sqrt{n/\sigma} \int_0^{\infty} \frac{1 - e^{-t^2}}{t^2}\, dt = \sqrt{\pi n/\sigma}$$

as $n \to \infty$, where $\tau(\epsilon) = \sqrt{n/\sigma}\, \alpha(\epsilon)$. Since $\beta(\epsilon) < 1/\alpha(\epsilon)$, it follows from (13.1) that

$$\frac{\sqrt{2}}{\sqrt{\pi(1 - \epsilon)}} \leq \liminf_{n\to\infty} \frac{I_n[|x|]}{\sqrt{n}} \leq \limsup_{n\to\infty} \frac{I_n[|x|]}{\sqrt{n}} \leq \sqrt{2/\pi}\,.$$

ϵ being arbitrary, this completes the proof.   □

SOLUTION 13.7

Put

$$f_n(x) = \frac{1}{n^{2+\epsilon}} \exp\left(\sqrt{\frac{2\log n}{n}} \left|\sum_{k=1}^{n} r_k(x)\right|\right).$$

It follows from **Problem 13.5** that

$$\int_0^1 f_n(x)\,dx < \frac{2}{n^{2+\epsilon}} \left(\frac{e^{\sqrt{(2\log n)/n}} + e^{-\sqrt{(2\log n)/n}}}{2} \right)^n$$

$$= \frac{2}{n^{1+\epsilon}} + O\left(n^{-3/2}\right)$$

as $n \to \infty$ for $0 < \epsilon < 1/2$; therefore

$$\sum_{n=1}^\infty \int_0^1 f_n(x)\,dx < \infty.$$

By Beppo Levi's theorem $\sum f_n(x)$ has a finite value almost everywhere. $\square$

Remark. In particular the above result implies that

$$\limsup_{n\to\infty} \frac{|r_1(x) + r_2(x) + \cdots + r_n(x)|}{\sqrt{n\log n}} \le \sqrt{2} \qquad (13.2)$$

almost everywhere.

To see this, note that almost every x in the interval $(0,1)$ satisfies

$$\frac{|r_1(x) + r_2(x) + \cdots + r_n(x)|}{\sqrt{n\log n}} < \frac{2+\epsilon}{\sqrt{2}}$$

for all sufficiently large n, since this inequality is equivalent to $f_n(x) < 1$. Now for any positive integer k let E_k be the set of points $x \in (0,1)$ such that the superior limit of

$$\frac{|r_1(x) + r_2(x) + \cdots + r_n(x)|}{\sqrt{n\log n}},$$

as $n \to \infty$, exceeds $\sqrt{2} + 1/k$. The above result implies that the set E_k is a null set for every k and so is the infinite union

$$\bigcup_{k=1}^\infty E_k.$$

The inequality (13.2) was first shown by Hardy and Littlewood (1914) in a different way. Later Khintchine (1924) strengthened (13.2) to

$$\limsup_{n\to\infty} \frac{|r_1(x) + r_2(x) + \cdots + r_n(x)|}{\sqrt{n\log\log n}} = \sqrt{2}$$

almost everywhere, which is called the law of the iterated logarithm.

Chapter 14

Legendre Polynomials

A. M. Legendre (1752–1833) was the first who introduced an orthogonal system of polynomials with weight function 1 on the interval $[-1, 1]$ in 1784 in his papers on celestial mechanics. We have already encountered this polynomial in **PROBLEM 4.6** with

$$P_n(x) = \frac{1}{\pi} \int_0^\pi \left(x + i\sqrt{1 - x^2} \cos\theta \right)^n d\theta,$$

known as the Laplace-Mehler integral.

- O. Rodrigues (1794–1851) gave the formula:

$$P_n(x) = \frac{1}{2^n n!} \frac{d^n}{dx^n} (x^2 - 1)^n.$$

The nth Legendre polynomial $P_n(x)$ is a degree n polynomial with rational coefficients. The first seven polynomials are as follows:

$$P_0(x) = 1,$$
$$P_1(x) = x,$$
$$P_2(x) = \frac{1}{2}(3x^2 - 1),$$
$$P_3(x) = \frac{1}{2}(5x^3 - 3x),$$
$$P_4(x) = \frac{1}{8}(35x^4 - 30x^2 + 3),$$
$$P_5(x) = \frac{1}{8}(63x^5 - 70x^3 + 15x),$$
$$P_6(x) = \frac{1}{16}(231x^6 - 315x^4 + 105x^2 - 5).$$

It follows from the Rodrigues formula that

$$P_n(x) = \frac{1}{2^n n!} \sum_{k=0}^{[n/2]} (-1)^k \binom{n}{k} \frac{(2n-2k)!}{(n-2k)!} x^{n-2k}.$$

Obviously $P_n(x)$ is even or odd according to the parity of n.

- Let $\delta_{n,m}$ be the Kronecker delta; namely, $\delta_{j,k} = 0$ for any $j \neq k$ and $\delta_{k,k} = 1$. We have

$$\int_{-1}^{1} P_n(x)P_m(x)\,dx = \frac{2}{2n+1}\,\delta_{n,m}.$$

- [*Recursion Formula*]

$$(n+1)P_{n+1}(x) = (2n+1)xP_n(x) - nP_{n-1}(x).$$

- [*Differential Equation*]

$$(1-x^2)P_n''(x) - 2xP_n'(x) + n(n+1)P_n(x) = 0.$$

PROBLEM 14.1

Show that

$$\int_{-1}^{1} (P_n'(x))^2 \, dx = n(n+1).$$

PROBLEM 14.2

Show that

$$\frac{1}{\sqrt{1-2xy+x^2}} = \sum_{n=0}^{\infty} P_n(y)x^n$$

for any $|y| \leq 1$ and $|x| < 1$.

Compare with PROBLEM 15.1.

PROBLEM 14.3

Show that

$$|P_n(\cos\varphi)| < \frac{c}{\sqrt{n}\,\sin\varphi}$$

for any $0 < \varphi < \pi$ with $c = \sqrt{\pi/2}$.

This estimate is not sharp at all. It was Stieltjes (1890) who showed that

$$|P_n(\cos\varphi)| < \frac{c'}{\sqrt{n}\,\sin\varphi}$$

for any $0 < \varphi < \pi$ with some constant c'. Gronwall (1913) gave a proof with $c' = 2\sqrt{2/\pi}$. Fejér (1925) gave a worse inequality with $c' = 4\sqrt{2/\pi}$ but in an elementary way. Finally Bernstein (1931) obtained the inequality with $c' = \sqrt{2/\pi}$, which cannot be replaced by any smaller constant since

$$P_{2n}(0) = \frac{(-1)^n}{2^{2n}}\binom{2n}{n} \sim \frac{(-1)^n}{\sqrt{\pi n}}$$

as $n \to \infty$. Bernstein's proof will be shown in the remark after **SOLUTION 14.3**.

PROBLEM 14.4

For any $0 < \varphi < \pi$ show that

$$P_n(\cos\varphi) = \frac{\sqrt{2}}{\pi}\int_0^\varphi \frac{\cos(n+1/2)\theta}{\sqrt{\cos\theta - \cos\varphi}}\,d\theta.$$

This is due to Mehler (1872), known as the Dirichlet-Mehler integral.

PROBLEM 14.5

Show that

$$P_n^2(x) \geq P_{n+1}(x)P_{n-1}(x)$$

for any $|x| \leq 1$ with equality only at $x = \pm 1$.

This is due to P. Turán (1910–1976). Szegö (1948) gave four different proofs for Turán's inequality. Later Turán published his original proof in Turán (1950).

PROBLEM 14.6

Show that

$$(P'_n(x))^2 > P'_{n+1}(x) P'_{n-1}(x)$$

for any $|x| \le 1$.

PROBLEM 14.7

Suppose that $\{p_n(x)\}_{n \ge 0}$ is a system of orthogonal polynomials on the interval $[a, b]$ with positive integrable weight $\rho(x)$; namely, $\deg p_n = n$ and

$$\int_a^b p_n(x) p_m(x) \rho(x) dx = 0$$

for any $n \ne m$. Show that each $p_n(x)$ possesses n simple roots in the interval (a, b) and that there exists an exactly one root of $p_{n-1}(x)$ between any consecutive roots of $p_n(x)$. Note that $p_0(x)$ is a non-zero constant.

This is one of various properties satisfied by general orthogonal polynomials. For details see the book of Szegö (1934).

PROBLEM 14.8

Show that

$$(n - 1)n(n + 1)(n + 2) \int_1^x \int_1^t P_n(s) \, ds \, dt = (1 - x^2)^2 P''_n(x).$$

Solutions for Chapter 14

SOLUTION 14.1

Integrating by parts we have

$$\int_{-1}^{1} (P_n'(x))^2 \, dx = \Big[P_n(x) P_n'(x) \Big]_{x=-1}^{x=1} - \int_{-1}^{1} P_n(x) P_n''(x) \, dx$$
$$= P_n'(1) - (-1)^n P_n'(-1),$$

since $P_n(1) = 1$ and $P_n(-1) = (-1)^n$. Note that the nth Legendre polynomial $P_n(x)$ is orthogonal to any polynomial of degree less than n. To determine the value of the derivative we use Leibniz' rule so that

$$P_n'(x) = \frac{1}{2^n n!} \frac{d^{n+1}}{dx^{n+1}} \big((x+1)^n (x-1)^n \big)$$
$$= \frac{1}{2^n} \sum_{k=1}^{n} \binom{n+1}{k} \binom{n}{k} k (x+1)^{k-1} (x-1)^{n-k};$$

therefore

$$P_n'(1) = \frac{n(n+1)}{2} \quad \text{and} \quad P_n'(-1) = (-1)^{n-1} \frac{n(n+1)}{2},$$

which implies

$$\int_{-1}^{1} (P_n'(x))^2 \, dx = n(n+1).$$

$\square$

SOLUTION 14.2

We first notice that $|P_n(y)| \le 1$ for any $|y| \le 1$ by the Laplace-Mehler integral. Hence the radius of convergence of the series on the right-hand side is greater than or equal to 1 as a power series of complex variable x. Thus it suffices to show the given formula for any sufficiently small $x > 0$.

Let $G(x)$ be the generating function of the Legendre polynomials; that is,

$$G(x) = \sum_{n=0}^{\infty} P_n(y) x^n.$$

Applying Cauchy's integral formula to the Rodrigues formula we obtain

$$G(x) = \sum_{n=0}^{\infty} \frac{1}{2\pi i} \int_{C_y} \left(\frac{x(z^2-1)}{2(z-y)} \right)^n \frac{dz}{z-y}$$

$$= -\frac{1}{\pi i} \int_{C_y} \frac{dz}{xz^2 - 2z + 2y - x},$$

where C_y is the oriented unit circle centered at y, that is, $C_y = \{z \in \mathbb{C} \; ; |z-y| = 1\}$. Let $\phi(z - y)$ be the denominator of the integrand of the last integral; namely,

$$\phi(w) = xw^2 - 2(1 - xy)w - x(1 - y^2),$$

which is a quadratic polynomial with real coefficients and it is easily verified that $\phi(w)$ has two real roots $\alpha_\pm$ in the intervals $(1, \infty)$ and $(-1, 0)$ respectively for all sufficiently small $x > 0$. Thus, if we put

$$z_\pm = y + \alpha_\pm = \frac{1 \pm \sqrt{1 - 2xy + x^2}}{x}$$

respectively, z_+ lies outside of C_y and z_- lies inside of C_y. Since

$$\phi(z - y) = x(z - z_+)(z - z_-),$$

it follows from the residue theorem that

$$G(x) = -\frac{2}{x(z_- - z_+)} = \frac{1}{\sqrt{1 - 2xy + x^2}}.$$

$\square$

$\boxed{\text{SOLUTION } 14.3}$

It follows from the Laplace-Mehler integral that

$$|P_n(\cos\varphi)| \le \frac{2}{\pi} \int_0^{\pi/2} \left(1 - \sin^2\varphi \sin^2\theta\right)^{n/2} d\theta.$$

Using Jordan's inequality $\sin\theta \ge 2\theta/\pi$, the right-hand side is estimated from above by

$$\frac{2}{\pi} \int_0^{\pi/2} \left(1 - \frac{4\theta^2}{\pi^2} \sin^2\varphi\right)^{n/2} d\theta,$$

and by the substitution $t = 2\theta/\pi$,

$$\int_0^1 \left(1 - t^2 \sin^2\varphi\right)^{n/2} dt.$$

Moreover using the inequality $1 - s \le e^{-s}$, it is less than or equal to

$$\int_0^1 \exp\left(-\frac{n}{2}t^2 \sin^2\varphi\right)dt.$$

Finally by the substitution $\tau = \sqrt{n/2}\,(\sin\varphi)t$, we obtain

$$|P_n(\cos\varphi)| < \int_0^\infty \exp\left(-\frac{n}{2}t^2\sin^2\varphi\right)dt$$

$$= \frac{\sqrt{2/n}}{\sin\varphi}\int_0^\infty e^{-\tau^2}\,d\tau.$$

Since the integral on the right-hand side is equal to $\sqrt{\pi}/2$, we obtain

$$|P_n(\cos\varphi)| < \frac{\sqrt{\pi/2}}{\sqrt{n}\,\sin\varphi}.$$

$\square$

REMARK. Bernstein's tricky proof is as follows. Put

$$f(\theta) = \sqrt{\sin\theta}\,P_n(\cos\theta)$$

for $0 < \theta < \pi$, which satisfies the differential equation $f''(\theta) + A(\theta)f(\theta) = 0$ with

$$A(\theta) = \frac{1}{4\sin^2\theta} + \left(n + \frac{1}{2}\right)^2.$$

We then put

$$F(\theta) = f^2(\theta) + \frac{(f'(\theta))^2}{A(\theta)};$$

hence

$$F'(\theta) = -\left(\frac{f'(\theta)}{A(\theta)}\right)^2 A'(\theta).$$

Since

$$2A'(\theta) = -\frac{1}{2}\cdot\frac{\cos\theta}{\sin^3\theta},$$

$F(\theta)$ is monotone increasing on $(0, \pi/2]$ and decreasing on $[\pi/2, \pi)$. Moreover since $f(0+) = 0$ and $|F(\theta)| = |F(\pi - \theta)|$, it follows that

$$f^2(\theta) \le F(\theta) \le F\left(\frac{\pi}{2}\right) = P_n^2(0) + \frac{(P_n'(0))^2}{n^2 + n + 1/2}.$$

If $n = 2m$, then $P'_n(0) = 0$ and $|f(\theta)| \le |P_n(0)|$. If $n = 2m + 1$, then $P_n(0) = 0$ and

$$|f(\theta)| < \frac{|P'_n(0)|}{\sqrt{n^2 + n + 1/2}} = \frac{n|P_{n-1}(0)|}{\sqrt{n^2 + n + 1/2}}.$$

Now it follows from the Rodrigues formula or the Laplace-Mehler integral that

$$|P_{2m}(0)| = \frac{1}{2^{2m}} \binom{2m}{m},$$

say c_m. The desired inequality follows from the following properties on c_m:

$$\sqrt{2m}\, c_m \quad \text{and} \quad \frac{(2m + 1)^{3/2}}{\sqrt{4m^2 + 6m + 5/2}}\, c_m$$

are monotone increasing sequences converging to $\sqrt{2/\pi}$ as $m \to \infty$. This limit may be obtained by Stirling's approximation mentioned after **Solution 16.7**.

Solution 14.4

For an arbitrary fixed $x \in (-1, 1)$ the point

$$z = x + i\sqrt{1 - x^2}\, \cos\theta$$

moves steadily on the segment AB downward, as θ varies from 0 to π, where

$$A = x + i\sqrt{1 - x^2} \quad \text{and} \quad B = x - i\sqrt{1 - x^2}.$$

Applying this transformation to the Laplace-Mehler integral, we get

$$\begin{aligned}
P_n(x) &= \frac{1}{\pi} \int_{AB} z^n \frac{d\theta}{dz}\, dz \\
&= \frac{i}{\pi} \int_{AB} \frac{z^n}{\sqrt{1 - x^2}\, \sin\theta}\, dz \\
&= \frac{i}{\pi} \int_{AB} \frac{z^n}{\sqrt{1 - 2xz + z^2}}\, dz,
\end{aligned}$$

where the square root is determined in such a way that the real part is positive; namely, the function $\sqrt{w}$ maps the wedge on the full angle by exclusion of the negative real axis onto the right half plane. Hence, $\sqrt{1 - 2xz + z^2}$ is analytic on the region Ω by exclusion of the two vertical half lines defined by

$$\ell^{\pm} = \left\{ z \in \mathbb{C} ; z = x \pm it,\ t \ge \sqrt{1 - x^2} \right\}.$$

Now we can change the segment AB by the circular arc through the point $z = 1$ joining A and B. If $x = \cos \varphi$ for $0 < \varphi < \pi$, this arc is expressed as $z = e^{i\theta}$ for $\varphi \geq \theta \geq -\varphi$; hence

$$P_n(\cos \varphi) = \frac{1}{\pi} \int_{-\varphi}^{\varphi} \frac{e^{i(n+1)\theta}}{\sqrt{1 - 2e^{i\theta} \cos \varphi + e^{2i\theta}}} \, d\theta.$$

Since

$$1 - 2e^{i\theta} \cos \varphi + e^{2i\theta} = 2(\cos \theta - \cos \varphi) e^{i\theta},$$

we obtain

$$P_n(\cos \varphi) = \frac{\sqrt{2}}{\pi} \int_0^{\varphi} \frac{\cos(n + 1/2)\theta}{\sqrt{\cos \theta - \cos \varphi}} \, d\theta.$$

$\square$

SOLUTION 14.5

The proof goes on the lines of the fourth proof of Szegö (1934).
Since

$$\Delta_n(x) = P_n^2(x) - P_{n+1}(x) P_{n-1}(x)$$

is even, it suffices to consider the problem on the interval $[0, 1]$. Using the recursion formula satisfied by the Legendre polynomials, we get

$$\Delta_n(x) = A^2(x) + B(x) P_{n-1}^2(x)$$

where

$$\begin{cases} A(x) = P_n(x) - \dfrac{2n + 1}{2n + 2} x P_{n-1}(x), \\[2mm] B(x) = \dfrac{n}{n + 1} - \left(\dfrac{2n + 1}{2n + 2} x\right)^2. \end{cases}$$

Hence $\Delta_n(x)$ is clearly positive for

$$0 \leq x < \frac{\sqrt{n(n + 1)}}{n + 1/2}.$$

Moreover, if $P_{n+1}(\xi) = 0$, then $P_n(\xi) \neq 0$; otherwise, we would have $P_0(\xi) = 0$ by the recursion formula, a contradiction; so, $\Delta_n(\xi) > 0$.

Problems and Solutions in Real Analysis

We thus consider hereafter any point x in the interval

$$\left[\frac{\sqrt{n(n+1)}}{n+1/2}, 1 \right)$$

with $P_{n+1}(x) \neq 0$. Note that $\Delta_n(1) = 0$. We now introduce the following polynomial in y of degree $n+1$:

$$Q_{n+1}(y) = \sum_{k=0}^{n+1} \binom{n+1}{k} P_k(x) y^k.$$

It follows from the Laplace-Mehler integral that

$$Q_{n+1}(y) = \frac{1}{\pi} \int_0^\pi \left(1 + xy + iy\sqrt{1-x^2}\,\cos\theta \right)^{n+1} d\theta.$$

Solving the equation $\sigma^2(1+xy)^2 + (\sigma y)^2(1-x^2) = 1$ in σ, we get

$$\sigma = \frac{1}{\sqrt{1+2xy+y^2}};$$

therefore

$$Q_{n+1}(y) = \frac{1}{\sigma^{n+1}} P_{n+1}(\phi(y))$$

where

$$\phi(y) = \sigma(1+xy) = \frac{1+xy}{\sqrt{1+2xy+y^2}}$$

is strictly monotone increasing on $(-\infty, 0]$ and strictly monotone decreasing on $[0, \infty)$ with $\phi(-\infty) = -x$, $\phi(0) = 1$ and $\phi(\infty) = x$. This means that, since $P_{n+1}(x) \neq 0$, ϕ gives a one-to-one correspondence between the negative zeros of Q_{n+1} smaller than $\phi^{-1}(x) < 0$ and that of P_{n+1} lying on the interval $(-x, x)$. Similarly ϕ gives a two-to-one correspondence between the zeros of Q_{n+1} greater than $\phi^{-1}(x)$ and that of P_{n+1} lying on the interval $(x, 1)$. Since the zeros of P_{n+1} locate symmetrically in the interval $(-1, 1)$ with respect to the origin, it follows that $Q_{n+1}(y)$ has exactly $n+1$ simple real roots, say $y_1 < y_2 < \cdots < y_{n+1}$. Therefore, if we put

$$s_1 = \sum_{k=1}^{n+1} y_k = -\binom{n+1}{1} \frac{P_n(x)}{P_{n+1}(x)}$$

and

$$s_2 = \sum_{1 \le j < k \le n+1} y_j y_k = \binom{n+1}{2} \frac{P_{n-1}(x)}{P_{n+1}(x)},$$

then

$$s_1^2 - 2s_2 = \sum_{k=1}^{n+1} y_k^2 = \frac{1}{n} \sum_{1 \le j < k \le n+1} \left(y_j^2 + y_k^2 \right) > \frac{2}{n} s_2,$$

which implies $\Delta_n(x) > 0$.   □

SOLUTION 14.6

Let $\phi(x, y) = G(x)$ be the generating function for the Legendre polynomials as in SOLUTION 14.2; that is,

$$\phi(x, y) = \frac{1}{\sqrt{1 - 2xy + x^2}} = \sum_{n=0}^{\infty} P_n(y) x^n$$

for $|y| \le 1$ and $|x| < 1$. For brevity we put $A = \phi(xz, y)$ and $B = \phi(x/z, y)$ where z is a complex variable on the unit circle $|z| = 1$. Using

$$x \frac{\partial \phi}{\partial x} = x(y - x)\phi^3 = \sum_{n=1}^{\infty} nP_n(y) x^n$$

it follows from the residue theorem that

$$\sum_{n=1}^{\infty} n(n+1) P_n^2(y) x^{2n} = \frac{1}{2\pi i} \int_C \Phi_1 \frac{dz}{z},$$

where

$$\Phi_1 = xz(y - xz)\left(1 + \frac{x}{z}\left(y - \frac{x}{z}\right)B^2\right)A^3 B$$

$$= x(y - xz)(z - xy)A^3 B^3$$

and C is the unit circle $|z| = 1$. Similarly we have

$$\sum_{n=1}^{\infty} n(n+1) P_{n+1}(y) P_{n-1}(y) x^{2n} = \frac{1}{2\pi i} \int_C \Phi_2 \frac{dz}{z},$$

where

$$\Phi_2 = \frac{x}{z}(y - xz)\left(1 + \frac{x}{z}\left(y - \frac{x}{z}\right)B^2\right)A^3B$$
$$= \frac{x}{z^2}(y - xz)(z - xy)A^3B^3.$$

Therefore

$$\sum_{n=1}^{\infty} n(n+1)\Delta_n(y)x^{2n} = \frac{1}{2\pi i}\int_C \Phi_3 \frac{dz}{z},$$

where

$$\Phi_3 = \Phi_1 - \Phi_2 = x(y - xz)(z - xy)\left(1 - \frac{1}{z^2}\right)A^3B^3.$$

We now write $D(z) = x(y - xz)(z - xy)$. By the substitution $w = 1/z$ the unit circle $|z| = 1$ maps onto $|w| = 1$ in the opposite direction. This means that

$$\int_C \frac{dz}{z}$$

is invariant under this substitution, as well as AB, while $D(z)(1 - z^{-2})$ is transformed into

$$D\left(\frac{1}{z}\right)(1 - z^2) = -z^2 D\left(\frac{1}{z}\right)\left(1 - \frac{1}{z^2}\right).$$

Therefore

$$\int_C \Phi_3 \frac{dz}{z} = -\int_C z^2 D\left(\frac{1}{z}\right)\left(1 - \frac{1}{z^2}\right)A^3B^3 \frac{dz}{z}.$$

Since

$$D(z) - z^2 D\left(\frac{1}{z}\right) = x^2(1 - y^2)(1 - z^2),$$

we obtain

$$\int_C \Phi_3 \frac{dz}{z} = -\frac{1}{2}x^2(1 - y^2)\int_C \left(z - \frac{1}{z}\right)^2 A^3B^3 \frac{dz}{z}.$$

On the other hand, it is easily seen that

$$\frac{\partial \phi}{\partial y} = x\phi^3 = \sum_{n=1}^{\infty} P_n'(y)x^n,$$

which implies

$$x^2 A^3 B^3 = \sum_{n,m \geq 1} P'_n(y) P'_m(y) x^{n+m} z^{n-m}.$$

We thus have

$$\sum_{n=1}^{\infty} n(n+1) \Delta_n(y) x^{2n} = (1 - y^2) \sum_{n=1}^{\infty} E_n(y) x^{2n},$$

where

$$E_n(y) = (P'_n(y))^2 - P'_{n+1}(y) P'_{n-1}(y);$$

that is,

$$n(n+1) \Delta_n(y) = (1 - y^2) E_n(y)$$

for any positive integer n. Therefore $E_n(y) \geq 0$ for any $|y| \leq 1$ by virtue of Turán's inequality (**PROBLEM 14.5**), where the sign of equality is excluded since

$$E_n(\pm 1) = \frac{n(n+1)}{2}.$$

$\square$

SOLUTION 14.7

Suppose, on the contrary, that the number of distinct real roots in the interval (a, b) of $p_n(x)$ is less than n. Then the number of roots of odd order is clearly less than n, which implies that there exists a non-zero polynomial $q(x)$ of degree less than n satisfying $p_n(x) q(x) \geq 0$ on $[a, b]$, contrary to the orthogonality:

$$\int_a^b p_n(x) q(x) \rho(x) \, dx = 0.$$

Let α, β be any consecutive real roots of $p_n(x)$ and ℓ be the number of distinct roots of $p_{n-1}(x)$ lying in the interval $[\alpha, \beta]$. If $\ell = 0$, then $p_{n-1}(x)$ has the constant sign on $[\alpha, \beta]$; so, it is geometrically obvious that the curves $c p_n(x)$ and $p_{n-1}(x)$ are tangent together at some point in $[\alpha, \beta]$ for some constant $c \neq 0$. Similarly if $\ell \geq 2$, then we can choose a suitable constant c' so that the curves $p_n(x)$ and $c' p_{n-1}(x)$ are tangent together at some point in $[\gamma, \delta] \subset [\alpha, \beta]$ where γ, δ are any two consecutive roots of $p_{n-1}(x)$. This is possible even if $\alpha = \gamma$, since we can take $c' = p'_n(\alpha)/p'_{n-1}(\alpha)$ in that case. Anyway we can choose some constant c such that

$$p_n(x) - c p_{n-1}(x) = (x - \xi)^2 q(x)$$

for some ξ and some polynomial $q(x)$ of degree $n - 2$. Therefore

$$\int_a^b (p_n(x) - cp_{n-1}(x))q(x)\rho(x)\,dx = 0$$

by the orthogonality; however this is a contradiction since the integral on the left-hand side is equal to

$$\int_a^b (x - \xi)^2 q^2(x)\rho(x)\,dx.$$

Thus we have $\ell = 1$, as required. □

$\boxed{\text{SOLUTION } \textbf{14.8}}$

Using the differential equation

$$(1 - x^2)P_n'' = 2xP_n' - n(n + 1)P_n$$

we have

$$\begin{aligned}
\left((1 - x^2)^2 P_n''\right)' &= -2x(2xP_n' - n(n + 1)P_n) \\
&\quad + (1 - x^2)(2P_n' + 2xP_n'' - n(n + 1)P_n') \\
&= -(n - 1)(n + 2)(1 - x^2)P_n'.
\end{aligned}$$

Hence the derivative of the left-hand side is equal to

$$(n - 1)n(n + 1)(n + 2)P_n.$$

Therefore

$$(n - 1)n(n + 1)(n + 2)\int_1^x\int_1^t P_n(s)\,ds\,dt = (1 - x^2)^2 P_n''(x) + \alpha x + \beta$$

for some constants α and β. We thus have $\alpha = \beta = 0$ since the right-hand side should have the factor $(1 - x)^2$. □

Chapter 15

Chebyshev Polynomials

As is mentioned in the remark after **Solution 3.2**, the polynomial $T_n(x)$ defined by the relation

$$T_n(\cos \theta) = \cos n\theta$$

is called the nth Chebyshev polynomial of the first kind, first appeared in Chebyshev (1854). We encounter these polynomials in various extremal problems as well as in best approximation problems. The Chebyshev polynomials form an orthogonal system over the interval $[-1, 1]$ with respect to the measure

$$d\mu = \frac{dx}{\sqrt{1 - x^2}}.$$

Obviously $T_n(x)$ is even or odd according to the parity of n.

The nth Chebyshev polynomial $T_n(x)$ is a polynomial with integer coefficients of degree n. The first eight polynomials are as follows:

$$T_0(x) = 1,$$
$$T_1(x) = x,$$
$$T_2(x) = 2x^2 - 1,$$
$$T_3(x) = 4x^3 - 3x,$$
$$T_4(x) = 8x^4 - 8x^2 + 1,$$
$$T_5(x) = 16x^5 - 20x^3 + 5x,$$
$$T_6(x) = 32x^6 - 48x^4 + 18x^2 - 1,$$
$$T_7(x) = 64x^7 - 112x^5 + 56x^3 - 7x.$$

Problems and Solutions in Real Analysis

- Let $\delta_{n,m}$ be the Kronecker delta. We have

$$\int_{-1}^{1} T_n(x) T_m(x) \, d\mu = \frac{\pi}{2} \delta_{n,m}$$

except for $(n, m) \ne (0, 0)$. If $n = m = 0$, the corresponding value is π.

- $T_n(T_m(x)) = T_{nm}(x)$ and $2T_n(x) T_m(x) = T_{n+m}(x) + T_{|n-m|}(x)$ for any n and m.

- [*Recursion Formula*]

$$T_{n+1}(x) = 2x T_n(x) - T_{n-1}(x).$$

- [*Differential Equation*]

$$(1 - x^2) T_n''(x) - x T_n'(x) + n^2 T_n(x) = 0.$$

The nth Chebyshev polynomial of the second kind $U_n(x)$ defined by

$$U_n(\cos \theta) = \frac{\sin(n + 1)\theta}{\sin \theta}$$

appeared in the remark after **Solution 5.6**. It satisfies the same recursion formula as T_n, as an independent solution of the recursion formula.

Problem 15.1

Show that

$$\frac{1 - xy}{1 - 2xy + x^2} = \sum_{n=0}^{\infty} T_n(y) x^n$$

for any $|y| \le 1$ and $|x| < 1$.

Compare with **Problem 14.2**.

Problem 15.2

Show that

$$T_n(x) = \frac{1}{2} \left(\left(x + \sqrt{x^2 - 1} \, \right)^n + \left(x - \sqrt{x^2 - 1} \, \right)^n \right)$$

for any $|x| > 1$.

PROBLEM 15.3

Show that

$$T_n(x) = \sum_{k=0}^{[n/2]} (-1)^k \frac{n}{n-k} \binom{n-k}{k} 2^{n-2k-1} x^{n-2k}.$$

PROBLEM 15.4

Show the following Rodrigues formula:

$$T_n(x) = \frac{(-1)^n}{1 \cdot 3 \cdots (2n-1)} \sqrt{1-x^2} \, \frac{d^n}{dx^n} \left(1-x^2\right)^{n-1/2}.$$

PROBLEM 15.5

Let $Q(x)$ be any polynomial of degree $\leq n$ with real coefficients and let M be the maximum of $|Q(x)|$ on the interval $[-1, 1]$. Show then that

$$|Q(x)| \leq M|T_n(x)|$$

for any $|x| > 1$.

This is due to Chebyshev (1881).

PROBLEM 15.6

Let $Q(x)$ be any monic polynomial of degree n with real coefficients. (A polynomial is said to be monic if its leading coefficient is unity.) Show then that

$$\max_{|x| \leq 1} |Q(x)| \geq \frac{1}{2^{n-1}}$$

and the equality occurs if and only if $Q(x) = 2^{1-n} T_n(x)$.

PROBLEM 15.7

Show that

$$T_n^{(k)}(1) = \frac{n^2(n^2 - 1^2) \cdots (n^2 - (k-1)^2)}{1 \cdot 3 \cdot 5 \cdots (2k-1)}$$

for any $1 \leq k \leq n$.

208   *Problems and Solutions in Real Analysis*

PROBLEM 15.8

Show that

$$\frac{\pi}{2}\sqrt{1-x^2} = 1 - 2\sum_{n=1}^{\infty}\frac{T_{2n}(x)}{4n^2-1}.$$

Note that expanding a given function $f(x)$ on the interval $[-1, 1]$ into the Chebyshev series is nothing but expanding $f(\cos\theta)$ on the interval $[-\pi, \pi]$ into the cosine Fourier series.

PROBLEM 15.9

We use the same notations as in **PROBLEM 1.10** *for the interval* $E = [-1, 1]$. *Show that*

$$\frac{1}{2^{n-1}} \leq \frac{M_{n+1}}{M_n} \leq \frac{n+1}{2^{n-1}}.$$

Deduce from this that the transfinite diameter of $[-1, 1]$ *is equal to* $1/2$.

Solutions for Chapter 15

SOLUTION 15.1

The proof is much easier than the Legendre case (**PROBLEM 14.2**). For brevity put $y = \cos\theta$. Let $G(y)$ be the generating function of the Chebyshev polynomials; that is,

$$G(y) = \sum_{n=0}^{\infty} T_n(\cos\theta)x^n = \sum_{n=0}^{\infty} (\cos n\theta)x^n.$$

Since the right-hand side is the real part of the geometric series $\sum (e^{i\theta}x)^n$, it follows that

$$G(y) = \Re \frac{1}{1 - e^{i\theta}x} = \frac{1 - x\cos\theta}{1 + x^2 - 2x\cos\theta},$$

as required.   □

SOLUTION 15.2

Let

$$f_n(x) = \frac{1}{2}\left(\left(x + \sqrt{x^2 - 1}\right)^n + \left(x - \sqrt{x^2 - 1}\right)^n\right)$$

and

$$g_n(x) = \frac{\sqrt{x^2 - 1}}{2}\left(\left(x + \sqrt{x^2 - 1}\right)^n - \left(x - \sqrt{x^2 - 1}\right)^n\right).$$

Then it is easily seen that

$$\begin{pmatrix} f_{n+1}(x) \\ g_{n+1}(x) \end{pmatrix} = \begin{pmatrix} x & 1 \\ x^2 - 1 & x \end{pmatrix}\begin{pmatrix} f_n(x) \\ g_n(x) \end{pmatrix},$$

from which we can get the recursion formula

$$f_{n+1}(x) = 2x f_n(x) - f_{n-1}(x).$$

Since $f_0(x) = 1$ and $f_1(x) = x$, we have $f_n(x) = T_n(x)$ for all n.   □

$\boxed{\text{SOLUTION 15.3}}$

Putting $x = \cos\theta$ and expanding the right-hand side of

$$\cos n\theta = \Re\,(\cos\theta + i\sin\theta)^n,$$

we have

$$T_n(x) = \Re\sum_{k=0}^{n}\binom{n}{k}\cos^{n-k}\theta\,(i\sin\theta)^k$$

$$= \sum_{\ell=0}^{[n/2]}(-1)^\ell\binom{n}{2\ell}\cos^{n-2\ell}\theta\,(1-\cos^2\theta)^\ell.$$

Thus

$$T_n(x) = x^n - \binom{n}{2}x^{n-2}(1-x^2) + \binom{n}{4}x^{n-4}(1-x^2)^2 - \cdots,$$

which gives

$$a_{n-2k} = (-1)^k\sum_{\ell=k}^{[n/2]}\binom{n}{2\ell}\binom{\ell}{k}$$

where a_{n-2k} is the coefficient of x^{n-2k} in the nth Chebyshev polynomial $T_n(x)$. Since this can be written as

$$a_{n-2k} = \frac{(-1)^k}{k!}Q^{(k)}(1)$$

where

$$Q(x^2) = \sum_{\ell=0}^{[n/2]}\binom{n}{2\ell}x^{2\ell} = \frac{(1+x)^n + (1-x)^n}{2},$$

it follows from Cauchy's integral formula that

$$(-1)^k a_{n-2k} = \frac{1}{2\pi i}\int_{C_0}\frac{Q(1+z)}{z^{k+1}}\,dz$$

$$= \frac{1}{4\pi i}\int_{C_0}\frac{\left(1+\sqrt{1+z}\right)^n + \left(1-\sqrt{1+z}\right)^n}{z^{k+1}}\,dz,$$

where C_0 is a circle centered at the origin with a small radius and the square root $\sqrt{1+z}$ is determined as the real part is positive. Since $1 - \sqrt{1+z} = O(|z|)$ in a neighborhood of the origin, we have

$$a_{n-2k} = \frac{(-1)^k}{4\pi i} \int_{C_0} \frac{\left(1 + \sqrt{1+z}\right)^n}{z^{k+1}} \, dz.$$

Therefore, by the substitution $w = \sqrt{1+z}$, we see from above that $(-1)^k a_{n-2k}$ is equal to

$$\frac{1}{2\pi i} \int_{C_0} \frac{w(1+w)^n}{(w^2-1)^{k+1}} \, dw = \frac{1}{2\pi i} \int_{C_1} \frac{w(1+w)^{n-k-1}}{(w-1)^{k+1}} \, dw$$

$$= \frac{1}{2\pi i} \int_{C_2} \frac{(1+\zeta)(2+\zeta)^{n-k-1}}{\zeta^{k+1}} \, d\zeta,$$

where C_1 and C_2 are circles centered at $w = 1$ and $\zeta = 0$ respectively with small radii. The last integral is clearly equal to

$$\binom{n-k-1}{k} 2^{n-2k-1} + \binom{n-k-1}{k-1} 2^{n-2k}$$

$$= \frac{n}{n-k} \binom{n-k}{k} 2^{n-2k-1}.$$

$\square$

SOLUTION 15.4

Put

$$Q(x) = \sqrt{1-x^2} \, \frac{d^n}{dx^n} \left(1 - x^2\right)^{n-1/2}.$$

It follows from Leibniz' rule that

$$Q(x) = \sqrt{1-x^2} \sum_{k=0}^{n} \binom{n}{k} \left((1+x)^{n-1/2}\right)^{(n-k)} \left((1-x)^{n-1/2}\right)^{(k)}.$$

The right-hand side can be written as

$$\sum_{k=0}^{n} (-1)^k \binom{n}{k} \frac{(2n-1)\cdots(2k+1)}{2^{n-k}} (1+x)^k$$

$$\times \frac{(2n-1)\cdots(2n-2k+1)}{2^k} (1-x)^{n-k},$$

Problems and Solutions in Real Analysis

which implies that $Q(x)$ is a polynomial of degree $\le n$. Indeed the coefficient of x^n in $Q(x)$ is equal to

$$\frac{(-1)^n}{2^n} \sum_{k=0}^{n} \binom{n}{k} (2n-1)\cdots(2k+1) \times (2n-1)\cdots(2n-2k+1),$$

which does not vanish. The reader may evaluate this sum as

$$\frac{(-1)^n}{2} \cdot \frac{(2n)!}{n!},$$

using the expansion of

$$\frac{1}{2}\Big((1+x)^{2n} + (1-x)^{2n}\Big).$$

On the other hand, integrating by parts for $k+1$ times, we get

$$\int_{-1}^{1} x^k Q(x)\, d\mu = (-1)^{k+1} \int_{-1}^{1} \left(x^k\right)^{(k+1)} \frac{d^{n-k-1}}{dx^{n-k-1}} \left(1-x^2\right)^{n-1/2} dx$$

$$= 0$$

for any non-negative integer $k < n$. Since the degree of the polynomial

$$R(x) = T_n(x) - \frac{(-2)^n n!}{(2n)!} Q(x)$$

is less than n and satisfies

$$\int_{-1}^{1} x^k R(x)\, d\mu = 0$$

for any $0 \le k < n$, we have

$$\int_{-1}^{1} R^2(x)\, d\mu = 0;$$

hence $R(x)$ vanishes everywhere. In other words,

$$T_n(x) = \frac{(-2)^n n!}{(2n)!} Q(x) = \frac{(-1)^n}{1\cdot 3 \cdots (2n-1)} Q(x).$$

$\square$

SOLUTION 15.5

Suppose, contrary to the conclusion, that $|Q(x_0)| > M|T_n(x_0)|$ for some point x_0 satisfying $|x_0| > 1$. Put $c = Q(x_0)/T_n(x_0)$ for brevity. Consider the polynomial

$$R(x) = cT_n(x) - Q(x)$$

of degree $\leq n$. If we put

$$\alpha_k = \cos\frac{k\pi}{n}$$

for $0 \leq k \leq n$, then clearly $T_n(\alpha_k) = (-1)^k$; therefore,

$$\operatorname{sgn} R(\alpha_k) = (-1)^k \operatorname{sgn} c,$$

since $|c| > M$ and $|Q(\alpha_k)| \leq M$. This implies that the polynomial R vanishes at least at n points in the interval $(-1, 1)$. But then, since $R(x_0) = 0$, R must vanish at least $n + 1$ points, a contradiction. $\qquad\square$

SOLUTION 15.6

Since

$$Q(\cos\theta) = \frac{\cos n\theta}{2^{n-1}},$$

it is easily seen that the absolute value of

$$Q(x) = \frac{T_n(x)}{2^{n-1}} = x^n + \cdots$$

attains its maximum 2^{1-n} at $y_k = \cos(k\pi/n)$ for each $0 \leq k \leq n$. Suppose now that there exists a polynomial $R(x) = x^n + \cdots$ satisfying

$$\max_{|x|\leq 1} |R(x)| < \frac{1}{2^{n-1}}.$$

Then clearly we have $R(y_0) < Q(y_0)$, $R(y_1) > Q(y_1)$, ... so that the polynomial $R(x) - Q(x)$ has at least one zero point in each interval (y_{k+1}, y_k). Hence $R(x) - Q(x)$ has at least n zero points in the interval $(-1, 1)$, contrary to the fact that the degree of $R(x) - Q(x)$ is less than n.

Next let $U(x) = x^n + \cdots$ be any polynomial with real coefficients satisfying

$$\max_{|x|\leq 1} |U(x)| = \frac{1}{2^{n-1}}.$$

Let m be the number of points $x \in [-1, 1]$ satisfying $|U(x)| = 2^{1-n}$, which are denoted by $x_1 < \cdots < x_m$. We then have $m = n + 1$. To see this it suffices to show $m > n$, because $U'(x_k) = 0$ for $1 < k < m$. Suppose, on the contrary, that $m \le n$. For any consecutive points x_i and x_{i+1} satisfying

$$\operatorname{sgn} U(x_i) \operatorname{sgn} U(x_{i+1}) = -1$$

there exists at least one zero point ξ of U in the interval (x_i, x_{i+1}). We take only one such a zero point for each (x_i, x_{i+1}) and name them $\xi_1 < \cdots < \xi_M$. Clearly

$$M \le m - 1 \le n - 1.$$

Now we define the polynomial

$$V(x) = c(x - \xi_1) \cdots (x - \xi_M)$$

where $c = \pm 1$ is chosen so that the sign of V on the interval (ξ_i, ξ_{i+1}) coincides with $\operatorname{sgn} U(x_k)$ for some $x_k \in (\xi_i, \xi_{i+1})$. In each interval (ξ_i, ξ_{i+1}) where V is positive, U may take a negative value somewhere in this interval. If so, the local minimum of U on (ξ_i, ξ_{i+1}) is certainly greater than -2^{1-n}. This means that the local maximum of the absolute value of $U(x) - \epsilon V(x)$ on this interval is less than 2^{1-n} for any sufficiently small $\epsilon > 0$. Moreover this argument is valid for the intervals $(-1, \xi_1)$ and $(\xi_M, 1)$. Therefore

$$\max_{|x| \le 1} |U(x) - \epsilon V(x)| < \frac{1}{2^{n-1}}.$$

Since the degree of V is less than n, this is contrary to the previous result. Hence we have $m = n + 1$.

Since

$$\max_{|x| \le 1} \left| \frac{Q(x) + U(x)}{2} \right| \le \frac{1}{2} \max_{|x| \le 1} |Q(x)| + \frac{1}{2} \max_{|x| \le 1} |U(x)|$$
$$= \frac{1}{2^{n-1}},$$

there exist $n + 1$ points $w_1 < \cdots < w_{n+1}$ in the interval $[-1, 1]$ satisfying

$$|Q(w_k) + U(w_k)| = \frac{1}{2^{n-2}}$$

by the same argument as above to the polynomial $(Q(x) + U(x))/2$. Thus we obtain

$$Q(w_k) = U(w_k) = \pm \frac{1}{2^{n-1}}$$

for any $1 \le k \le n + 1$, which implies that U coincides with Q.   □

SOLUTION 15.7

It follows from k times differentiation of the recursion formula that

$$\sum_{\ell=0}^{2} \binom{k}{\ell} (1 - x^2)^{(\ell)} T_n^{(k+2-\ell)}(x)$$

$$- \sum_{j=0}^{1} \binom{k}{\ell} x^{(\ell)} T_n^{(k+1-\ell)}(x) + n^2 T_n^{(k)}(x) = 0,$$

which implies that

$$T_n^{(k+1)}(1) = \frac{n^2 - k^2}{2k + 1} T_n^{(k)}(1).$$

One can deduce the desired formula using this repeatedly, in view of $T_n(1) = 1$.   □

SOLUTION 15.8

Put

$$\phi(x) = \frac{\pi}{2} \sqrt{1 - x^2} - 1 + 2 \sum_{n=1}^{\infty} \frac{T_{2n}(x)}{4n^2 - 1}.$$

Since $|T_n(x)| \le 1$ for $-1 \le x \le 1$, the Chebyshev series on the right-hand side converges uniformly. Thus we can employ the termwise integration to obtain

$$\int_{-1}^{1} T_{2m}(x)\phi(x) \, d\mu = \frac{\pi}{2} \int_{-1}^{1} T_{2m}(x) \, dx - \int_{-1}^{1} T_{2m}(x) \, d\mu$$

$$+ 2 \sum_{n=1}^{\infty} \frac{1}{4n^2 - 1} \int_{-1}^{1} T_{2m}(x) T_{2n}(x) \, d\mu$$

for any non-negative integer m. By the substitution $x = \cos\theta$ the first integral on the right-hand side is transformed into

$$\frac{\pi}{2} \int_{0}^{\pi} \cos 2m\theta \sin\theta \, d\theta,$$

which is equal to $-\pi/(4m^2 - 1)$. By the orthogonality the second integral vanishes except for $m = 0$, in which it is equal to π. Similarly the third integral vanishes except for $n = m$ and is equal to

$$\frac{\pi}{4m^2 - 1}$$

when $n = m \geq 1$. Hence ϕ is orthogonal to every even Chebyshev polynomial with respect to $d\mu$. But it is also orthogonal to every odd Chebyshev polynomial because ϕ is even. Therefore $\phi(x)$ vanishes everywhere by the remark after **PROBLEM 8.5**. □

SOLUTION 15.9

Let $\xi_1, ..., \xi_n$ be the points in the interval $[-1, 1]$ at which $|V(x_1, ..., x_n)|$ attains its maximum M_n and let ξ_0 be the point in $[-1, 1]$ at which the absolute value of the polynomial

$$\phi(x) = (x - \xi_1) \cdots (x - \xi_n)$$

attains its maximum. Then

$$|\phi(\xi_0)| = \frac{|V(\xi_0, \xi_1, ..., \xi_n)|}{|V(\xi_1, ..., \xi_n)|} \leq \frac{M_{n+1}}{M_n};$$

hence it follows from **PROBLEM 15.6** that

$$\frac{1}{2^{n-1}} \leq \max_{|x| \leq 1} |\phi(x)| = |\phi(\xi_0)| \leq \frac{M_{n+1}}{M_n},$$

because the leading coefficient of $\phi(x)$ is unity.

On the other hand, let

$$Q(x) = \frac{1}{2^{n-1}} T_n(x) = x^n + \cdots$$

and $\eta_1, ..., \eta_{n+1}$ be the points at which $|V(x_1, ..., x_{n+1})|$ attains its maximum M_{n+1}. Then

$$V(\eta_1, ..., \eta_{n+1}) = \begin{vmatrix} 1 & \eta_1 & \cdots & \eta_1^{n-1} & \eta_1^n \\ \vdots & \vdots & \ddots & \vdots & \vdots \\ 1 & \eta_{n+1} & \cdots & \eta_{n+1}^{n-1} & \eta_{n+1}^n \end{vmatrix} = \begin{vmatrix} 1 & \eta_1 & \cdots & \eta_1^{n-1} & Q(\eta_1) \\ \vdots & \vdots & \ddots & \vdots & \vdots \\ 1 & \eta_{n+1} & \cdots & \eta_{n+1}^{n-1} & Q(\eta_{n+1}) \end{vmatrix}.$$

Using the expansion by cofactors by the last column we get

$$M_{n+1} \leq |Q(\eta_1)| \cdot |V(\eta_2, ..., \eta_{n+1})| + \cdots + |Q(\eta_{n+1})| \cdot |V(\eta_1, ..., \eta_n)|$$
$$\leq \frac{n+1}{2^{n-1}} M_n.$$

The above inequalities hold even for $n = 1$ if we define $M_1 = 1$.

Multiplying these inequalities we obtain

$$\frac{1}{2^{(n-1)(n-2)/2}} \leq M_n \leq \frac{n!}{2^{(n-1)(n-2)/2}},$$

which implies that $M_n^{2/(n(n-1))}$ converges to $1/2$ as $n \to \infty$. $\qquad \square$

Chapter 16

Gamma Function

- The function $\Gamma(s)$ defined by the improper integral

$$\Gamma(s) = \int_0^\infty x^{s-1} e^{-x} \, dx$$

  convergent for $s > 0$ is called the *Gamma function* and satisfies the functional relation

$$\Gamma(s + 1) = s\Gamma(s).$$

  It is also called the second Eulerian integral. It follows that $\Gamma(n + 1) = n!$ for any positive integer n. The reader should notice the shift of the argument in this formula.

- It is known that

$$\Gamma\left(\frac{1}{2}\right) = \sqrt{\pi}.$$

- The Gamma function is closely related to Euler's constant γ through

$$\Gamma'(1) = -\gamma.$$

   L. Euler (1707–1783) introduced the interpolation formula (**PROBLEM 16.1**) in a correspondence to C. Goldbach (1690–1764) in 1729 as a generalization of the factorial for the case s is rational. We owe to A. M. Legendre (1752–1833) its formulation in that form as well as the consideration for any $s > 0$, who also introduced the notation $\Gamma(s)$ and gave the names for two types of Euler's integrals.

- The function defined by the improper integral

$$B(s,t) = \int_0^1 x^{s-1}(1-x)^{t-1} \, dx$$

convergent for $s, t > 0$ is called the Beta function or the first Eulerian integral. The name 'Beta' was introduced for the first time in Binet (1839). The reason why this is the first is that Euler started his derivation with this integral. It is known that

$$B(s,t) = \frac{\Gamma(s)\Gamma(t)}{\Gamma(s+t)}.$$

For the proof see the former part of **SOLUTION 16.3**. This yields $\Gamma(1/2) = \sqrt{\pi}$ for $s = t = 1/2$.

For various topics about the Gamma function involving Hadamard's factorial function, as well as Euler's experimental derivation, see Davis (1959). There are also good elementary expositions about the Gamma function; for examples, Barnes (1899), Jensen (1916), Gronwall (1918), etc.

PROBLEM 16.1

For any positive s show that

$$\Gamma(s) = \lim_{n\to\infty} \frac{n^s n!}{s(s+1)\cdots(s+n)}.$$

PROBLEM 16.2

Show that the Gamma function $\Gamma(s)$ is logarithmically convex as well as convex. Show moreover that the Beta function $B(s,t)$ is logarithmically convex as well as convex with respect to s and t.

Gamma Function   221

PROBLEM 16.3 ───

Let Δ_{n-1} be the $(n-1)$-dimensional simplex defined by $x_1, x_2, ..., x_{n-1} \geq 0$ and $x_1 + \cdots + x_{n-1} \leq 1$ for each integer $n \geq 2$. For $s_1 > 0, ..., s_n > 0$ show that the $(n-1)$-dimensional integral

$$\int \cdots \int_{\Delta_{n-1}} x_1^{s_1-1} \cdots x_{n-1}^{s_{n-1}-1} (1 - x_1 - \cdots - x_{n-1})^{s_n-1} \, dx_1 \cdots dx_{n-1}$$

is equal to

$$\frac{\Gamma(s_1) \cdots \Gamma(s_n)}{\Gamma(s_1 + \cdots + s_n)}.$$

PROBLEM 16.4 ───

For any non-zero polynomial $P(x; z_0, z_1, ..., z_m)$ with $m + 2$ variables show that the Gamma function $\Gamma(x)$ does not satisfy the differential equation

$$P\left(x; y, y', y'', ..., y^{(m)}\right) = 0.$$

This was first shown by Hölder (1887) and another proof was given by Moore (1897). Barnes (1899), Ostrowski (1919) and Hausdorff (1925) gave simpler and shorter proofs. Ostrowski (1925) corrected a mistake in his earlier paper that was pointed out by Hausdorff and gave further shorter proof. Ostrowski and Moore used the functional relation

$$f(x + 1) = xf(x)$$

satisfied by the Gamma function $\Gamma(x)$, while Hölder, Barnes and Hausdorff used

$$\psi(x + 1) = \psi(x) + \frac{1}{x}$$

satisfied by the digamma function $\Gamma'(x)/\Gamma(x)$.

PROBLEM 16.5 ───

Suppose that $f(x) \in C(0, \infty)$ is positive, logarithmically convex and satisfies the functional equation

$$f(x + 1) = xf(x)$$

with $f(1) = 1$. Then show that $f(x) = \Gamma(x)$.

This was shown by Bohr and Mollerup (1922), and known as the Bohr-Mollerup theorem. Their proof was later simplified by Artin (1964).

PROBLEM 16.6

Show that

$$\frac{1}{\Gamma(s)} = s e^{\gamma s} \prod_{n=1}^{\infty} \left(1 + \frac{s}{n}\right) e^{-s/n}$$

where γ is Euler's constant.

This is known as Weierstrass' canonical product of order 1. This is valid for any complex number s since the convergence is uniform on compact sets in the whole complex plane. This was first found by Schlömilch (1844) but it was Weierstrass (1856) who established as the product theorem in the theory of functions, in which he used the notation $1/Fc(s)$ for the gamma function.

PROBLEM 16.7

Show that

$$\int_{x}^{x+1} \log \Gamma(s)\,ds = x(\log x - 1) + \frac{1}{2}\log(2\pi)$$

for any $x > 0$.

This is known as Raabe's integral, first shown by Raabe (1843) for any positive integer and subsequently (1844) for any positive real number x.

PROBLEM 16.8

Show that

$$\Gamma(s)\Gamma(1-s) = \frac{\pi}{\sin \pi s}$$

for any $0 < s < 1$.

This is known as Euler's reflection formula, which produces $\Gamma(1/2) = \sqrt{\pi}$ again.

PROBLEM 16.9 ──────────────────────────────

Using the Bohr-Mollerup theorem show that

$$\log \Gamma(s) = \int_0^\infty \left(s - 1 - \frac{1 - e^{-(s-1)x}}{1 - e^{-x}} \right) \frac{e^{-x}}{x} \, dx$$

for any $s > 0$.

This formula is due to Malmstén (1847). Cauchy (1841) had obtained this for positive integer s.

PROBLEM 16.10 ──────────────────────────────

Using Malmstén's formula in **PROBLEM 16.9** *show that*

$$\log \Gamma(s) = \left(s - \frac{1}{2} \right) \log s - s + \frac{1}{2} \log(2\pi) + \omega(s),$$

where

$$\omega(s) = \int_0^\infty \left(\frac{1}{2} - \frac{1}{x} + \frac{1}{e^x - 1} \right) \frac{e^{-sx}}{x} \, dx$$

for any $s > 0$.

This is known as Binet's first formula for the log Gamma function. The function $\omega(s)$ is the Laplace transform of the function

$$\frac{1}{x} \left(\frac{1}{2} - \frac{1}{x} + \frac{1}{e^x - 1} \right),$$

which is a monotone decreasing function in $C[0, \infty)$. Note that $\omega(s)$ gives the error term for Stirling's approximation (See the remark after **SOLUTION 16.7**). Binet (1839) also found another integral expression for $\omega(s)$, known as the second formula, as follows:

$$\omega(s) = 2 \int_0^\infty \frac{\arctan(x/s)}{e^{2\pi x} - 1} \, dx.$$

PROBLEM 16.11 ──────────────────────────────

Using the Fourier series of $\log \dfrac{\Gamma(s)}{\Gamma(1-s)}$ show that

$$\log \Gamma(s) + \frac{1}{2} \log \frac{\sin \pi s}{\pi} + (\gamma + \log 2\pi) \left(s - \frac{1}{2} \right) = \sum_{n=2}^\infty \frac{\log n}{n\pi} \sin 2n\pi s$$

for $0 < s < 1$, where γ is Euler's constant.

Problems and Solutions in Real Analysis

This is known as Kummer's series (1847), in which he used the following integral representation for Euler's constant:

$$\gamma = \int_0^\infty \left(e^{-x} - \frac{1}{1+x} \right) \frac{dx}{x}.$$

Note that Kummer's series at $s = 3/4$ yields

$$\frac{\log 3}{3} - \frac{\log 5}{5} + \frac{\log 7}{7} - \cdots = \pi \left(\frac{\gamma - \log \pi}{4} + \log \Gamma \left(\frac{3}{4} \right) \right),$$

while Hardy (1912) obtained

$$\frac{\log 2}{2} - \frac{\log 3}{3} + \frac{\log 4}{4} - \cdots = \frac{1}{2} \log^2 2 - \gamma \log 2.$$

Solutions for Chapter 16

SOLUTION 16.1

Putting

$$\Phi_n(s) = \frac{n^s n!}{s(s+1)\cdots(s+n)},$$

we have by the partial fraction expansion

$$\Phi_n(s) = n^s \sum_{k=0}^{n} (-1)^k \binom{n}{k} \frac{1}{s+k}$$

$$= n^s \int_0^1 x^{s-1}(1-x)^n\, dx$$

$$= \int_0^n t^{s-1}\left(1 - \frac{t}{n}\right)^n dt.$$

Let ϵ be any positive number satisfying $\epsilon(s+2) < 1$. The last integral can be written as $J_1 + J_2$, where J_1 and J_2 are the integrals over the interval $[0, n^\epsilon]$ and $[n^\epsilon, n]$ respectively.

For $0 \le t \le n^\epsilon$ we have

$$n \log\left(1 - \frac{t}{n}\right) = -t + O(n^{2\epsilon - 1});$$

therefore

$$J_1 = \int_0^{n^\epsilon} t^{s-1} e^{-t}\, dt + O(n^{\epsilon(s+2)-1})$$

as $n \to \infty$.

For $n^\epsilon \le t \le n$ there exists a positive constant c satisfying

$$n \log\left(1 - \frac{t}{n}\right) \le n \log\left(1 - n^{\epsilon-1}\right) \le -cn^\epsilon;$$

hence $J_2 = O(n^s \exp(-cn^\epsilon))$, which converges to 0 as $n \to \infty$. Thus $\Phi_n(s)$ converges to $\Gamma(s)$ as $n \to \infty$. $\qquad\square$

| SOLUTION 16.2 |

Note that we can differentiate in s repeatedly under the integral sign in the second Eulerian integral, since

$$\int_0^\infty (\log x)^\sigma x^{s-1} e^{-x} \, dx$$

converges uniformly on compact sets in s in the region $\sigma > 0$. Thus the convexity of $\Gamma(s)$ follows from

$$\int_0^\infty (\log x)^2 x^{s-1} e^{-x} \, dx > 0$$

and the logarithmic convexity follows from the fact that the quadratic function of y:

$$\int_0^\infty (\sigma + \log x)^2 x^{s-1} e^{-x} \, dx$$

is positive for all σ, since this implies $\Gamma'^2(s) < \Gamma(s)\Gamma''(s)$.

Similar argument can be applied to the first Eulerian integral for its convexity and its logarithmic convexity with respect to s and t. $\qquad\square$

REMARK. Moreover, since

$$\int_0^1 \big(\sigma \log x + \log(1-x)\big)^2 x^{s-1}(1-x)^{t-1} \, dx > 0,$$

we have

$$\left(\frac{\partial^2}{\partial s \partial t} B(s,t)\right)^2 < \frac{\partial^2}{\partial s^2} B(s,t) \frac{\partial^2}{\partial t^2} B(s,t).$$

In other words, the Beta function has the positive Hessian.

| SOLUTION 16.3 |

We first treat the case $n = 2$. For a fixed number $t > 0$ put

$$\Psi(s) = \frac{\Gamma(s+t)}{\Gamma(t)} B(s,t)$$

for $s > 0$. Obviously $\Psi(s) > 0$ and $\Psi(1) = 1$. Moreover we have

$$\Psi(s+1) = (s+t)\frac{\Gamma(s+t)}{\Gamma(t)} B(s+1,t)$$

$$= s\Psi(s),$$

since it follows by integration by parts that

$$B(s+1, t) = \frac{s}{t} B(s, t+1)$$

$$= \frac{s}{t} B(s, t) - \frac{s}{t} B(s+1, t).$$

As is shown in **SOLUTION 16.2**, both $\Gamma(s+t)$ and $B(s, t)$ are logarithmically convex with respect to s. This means that $\Psi(s)$ is logarithmically convex and it follows from **PROBLEM 16.5** that $\Psi(s) = \Gamma(s)$. The formula for $n = 2$ is thus proved. This method is due to Artin (1964).

The general case will be shown by induction on n. Suppose that the formula holds true for n. For an arbitrary fixed $(x_1, ..., x_{n-1})$ in the $(n-1)$-dimensional simplex Δ_{n-1} we consider the substitution

$$x_n = (1 - x_1 - x_2 - \cdots - x_{n-1})t$$

for $0 < t < 1$. Then the integral

$$\int \cdots \int_{\Delta_n} x_1^{s_1-1} \cdots x_n^{s_n-1} (1 - x_1 - \cdots - x_n)^{s_{n+1}-1} \, dx_1 \cdots dx_n$$

is transformed into

$$\int \cdots \int_{\Delta_{n-1}} x_1^{s_1-1} \cdots x_{n-1}^{s_{n-1}-1} (1 - x_1 - \cdots - x_{n-1})^{s_n+s_{n+1}-1} \, dx_1 \cdots dx_{n-1}$$

$$\times \int_0^1 t^{s_n-1} (1-t)^{s_{n+1}-1} \, dt$$

$$= \frac{\Gamma(s_1) \cdots \Gamma(s_{n-1}) \Gamma(s_n + s_{n+1})}{\Gamma(s_1 + \cdots + s_n + s_{n+1})} \cdot \frac{\Gamma(s_n) \Gamma(s_{n+1})}{\Gamma(s_n + s_{n+1})},$$

which shows the formula for $n + 1$.   □

SOLUTION 16.4

The proof is based on Ostrowski (1925). For any term of P in the form

$$A(x) z_0^{n_0} z_1^{n_1} \cdots z_m^{n_m},$$

where $A(x)$ is a polynomial only in x, we assign the index $(n_0, n_1, ..., n_m)$ and introduce a lexicographical order; namely, we say that $(n_0, n_1, ..., n_m)$ is higher than $(n_0', n_1', ..., n_m')$ if $n_m = n_m'$, ..., $n_{j+1} = n_{j+1}'$ and $n_j > n_j'$ for some $0 \leq j \leq m$. Clearly the indices form a totally ordered set. Note that we do not distinguish $(n_0, n_1, ..., n_m)$ from $(n_0, n_1, ..., n_m, 0, ..., 0)$.

For any polynomial $P(x; z_0, z_1, ..., z_m)$ satisfying

$$P(x; \Gamma, \Gamma', \Gamma'', ..., \Gamma^{(m)}) = 0 \qquad (16.1)$$

if exists, we assign the highest index $(n_0, n_1, ..., n_m)$ among the terms in the form discussed above, which is denoted by ind P. We then pick up a polynomial P^* having the lowest ind P among all polynomials P satisfying (16.1). Let

$$A^*(x) z_0^{v_0} z_1^{v_1} \cdots z_m^{v_m}$$

be the corresponding highest term with ind $P^* = (v_0, v_1, ..., v_m)$. We can also assume that $\deg A^*$ is the smallest among such polynomials and the leading coefficient is unity.

Let P be any polynomial satisfying (16.1) with ind $P = $ ind P^* and let $A(x)$ be the coefficient of the highest term of P. Put $A(x) = q(x) A^*(x) + r(x)$ with $\deg r < \deg A^*$. If $r \not\equiv 0$, the coefficient of the highest term of $P - q(x) P^*$ would be $r(x)$, contrary to the choice of A^*; hence $A(x) = q(x) A^*(x)$. If $P - q(x) P^* \not\equiv 0$, then the highest index of $P - q(x) P^*$ would be certainly lower than ind P^*, contrary to the choice of P^*. We thus have $P = q(x) P^*$.

Substituting $\Gamma(x + 1) = x\Gamma(x)$ in the differential equation (16.1) with $P = P^*$, we obtain a new equation

$$\begin{aligned} 0 &= P^*(x + 1; x\Gamma, x\Gamma' + \Gamma, ..., x\Gamma^{(m)} + m\Gamma^{(m-1)}) \\ &= Q(x; \Gamma, \Gamma', \Gamma'', ..., \Gamma^{(m)}), \quad \text{say.} \end{aligned}$$

The highest term of Q certainly comes from the expansion of

$$A^*(x + 1)(xz_0)^{v_0}(xz_1 + z_0)^{v_1} \cdots (xz_m + mz_{m-1})^{v_m};$$

therefore, ind $Q = $ ind P^* and the coefficient of the highest term of Q becomes $x^N A^*(x + 1)$ with $N = v_0 + v_1 + \cdots + v_m$. It follows from the above argument that

$$B(x) = \frac{x^N A^*(x + 1)}{A^*(x)}$$

is a polynomial of degree N and $Q = B(x) P^*$; hence

$$\begin{aligned} P^*(x + 1; xz_0, xz_1 + z_0, ..., xz_m + mz_{m-1}) \\ = B(x) P^*(x; z_0, z_1, ..., z_m). \end{aligned}$$

If $B(\alpha) = 0$ for some $\alpha \neq 0$, then $P^*(\alpha + 1; w_0, w_1, ..., w_m) \equiv 0$. Let

$$M = 1 + \max \left\{ \deg_{z_0} P^*, ..., \deg_{z_m} P^* \right\}$$

where $\deg_{z_k}$ means the degree with respect to z_k. Since an index $(n_0, n_1, ..., n_m)$ is higher than another index $(n'_0, n'_1, ..., n'_m)$ if and only if

$$n_0 + n_1 M + \cdots + n_m M^m > n'_0 + n'_1 M + \cdots + n'_m M^m,$$

it follows from

$$P^*(\alpha + 1; t, t^M, ..., t^{M^m}) \equiv 0$$

that the coefficient of every term of P^* vanishes at $\alpha + 1$, contrary to the choice of A^*. Therefore $B(x) = x^N$ and

$$P^*(x + 1; xz_0, xz_1 + z_0, ..., xz_m + mz_{m-1})$$
$$= x^N P^*(x; z_0, z_1, ..., z_m). \qquad (16.2)$$

Putting $x = 0$ in (16.2) we get

$$P^*(1; 0, z_0, ..., mz_{m-1}) \equiv 0,$$

which implies $R(1; w_1, ..., w_m) \equiv 0$ where

$$R(x; w_1, ..., w_m) = P^*(x; 0, w_1, ..., w_m);$$

hence R has the factor $x - 1$ by the same argument as above. Note that $R \not\equiv 0$; otherwise P^* would have the factor z_0, contrary to the choice of P^*. Similarly putting $x = 1$ and $z_0 = 0$ in (16.2) we get

$$P^*(2; 0, z_1, z_2 + z_1, ..., z_m + mz_{m-1}) \equiv 0,$$

which implies $R(2; w_1, ..., w_m) \equiv 0$; so, R has the factor $x - 2$. Repeating this argument we see that R has the factor $x - k$ for any positive integer k, a contradiction.

$\square$

SOLUTION 16.5

By the functional equation it is easily seen that $f(n) = (n - 1)!$ for any positive integer n. For any fixed x in the open interval $(0, 1)$ it follows from **PROBLEM 9.4** that

$$\log f(x + n + 1) \leq (1 - x) \log f(n + 1) + x \log f(n + 2)$$
$$= (1 - x) \log n! + x \log(n + 1)!;$$

namely, $f(x + n + 1) \leq (n + 1)^x n!$. Similarly it follows from

$$\log f(n + 1) \leq \frac{x}{1 + x} \log f(n) + \frac{1}{1 + x} \log f(x + n + 1)$$

that $n^x n! \leq f(x + n + 1)$. Hence, using $f(x + n + 1) = (x + n) \cdots (x + 1) x f(x)$,

$$\Phi_n(x) \leq f(x) \leq \left(1 + \frac{1}{n}\right)^x \Phi_n(x),$$

where

$$\Phi_n(x) = \frac{n^x n!}{x(x + 1) \cdots (x + n)}.$$

Since $\Phi_n(x)$ converges to $\Gamma(x)$ as $n \to \infty$ by **Problem 16.1**, we obtain $f(x) = \Gamma(x)$ as required.   □

SOLUTION 16.6

It follows from **Problem 16.1** that

$$\frac{1}{\Gamma(s)} = \lim_{n \to \infty} \frac{s(s + 1) \cdots (s + n)}{n^s n!}$$

$$= s \lim_{n \to \infty} \frac{(s + 1) \cdots (s + n)}{(n + 1)^s n!}.$$

The sequence on n inside of the limit sign can be written in the form

$$\prod_{k=1}^{n} \left(1 + \frac{s}{k}\right) \exp\left(s \log \frac{k}{k + 1}\right),$$

which is equal to the product of

$$\prod_{k=1}^{n} \left(1 + \frac{s}{k}\right) e^{-s/k}$$

and

$$\exp\left(s\left(1 + \frac{1}{2} + \cdots + \frac{1}{n} - \log(n + 1)\right)\right).$$

Obviously the last expression converges to $e^{\gamma s}$ where γ is Euler's constant.   □

SOLUTION 16.7

Put

$$f(x) = \int_x^{x+1} \log \Gamma(s) \, ds$$

for $x > 0$. Differentiation yields

$$f'(x) = \log \frac{\Gamma(x+1)}{\Gamma(x)} = \log x;$$

hence

$$f(x) = x(\log x - 1) + c$$

for some constant c. Note that $c = f(1) + 1$.

To determine the value of c we use the formula in **Problem 16.1** in the form

$$\log \Gamma(s) = -\log s + \lim_{n \to \infty} \left(s \log n + \log \frac{1}{s+1} + \cdots + \log \frac{n}{s+n} \right),$$

which can be written as

$$\log \Gamma(s+1) = \sum_{k=1}^{\infty} \left(s \log \frac{k+1}{k} - \log \frac{s+k}{k} \right),$$

and the series converges uniformly in $s \in [0, 1]$ in view of

$$s \log \frac{k+1}{k} - \log \frac{s+k}{k} = O\left(\frac{1}{k^2} \right).$$

Hence we may integrate the above expression termwise to obtain

$$f(1) = \int_0^1 \log \Gamma(s+1)\, ds$$

$$= \lim_{n \to \infty} \left(\frac{1}{2} \log n - (n+1)\log(n+1) + n + \log n! \right)$$

$$= -1 + \lim_{n \to \infty} A_n,$$

where

$$A_n = n + \log n! - \left(n + \frac{1}{2} \right) \log n.$$

Hence $c = \lim_{n \to \infty} A_n$, and so c is also the limit of the sequence

$$2A_n - A_{2n} = \log \frac{n!^2}{(2n)!} + \left(2n + \frac{1}{2} \right) \log 2 - \frac{1}{2} \log n,$$

which is equal to the logarithm of

$$\sqrt{2} \left(1 + \frac{1}{2n} \right) \frac{\sqrt{n}\, n!}{\frac{1}{2}\left(\frac{1}{2} + 1 \right) \cdots \left(\frac{1}{2} + n \right)}.$$

This converges to $\sqrt{2}\,\Gamma(1/2) = \sqrt{2\pi}$ as $n \to \infty$ by **PROBLEM 16.1**. Hence $c = \log\sqrt{2\pi}$. $\square$

REMARK. The above proof shows that the factorial $n!$ is asymptotically

$$\sqrt{2\pi n}\left(\frac{n}{e}\right)^n$$

as $n \to \infty$, known as *Stirling's approximation*. J. Stirling (1692–1770) gave this asymptotic formula for $n!$ in Methodus Differentialis (1730).

SOLUTION 16.8

It follows from **PROBLEM 16.1** that

$$\frac{1}{\Gamma(s)\Gamma(1-s)} = \lim_{n\to\infty} \frac{s(s+1)\cdots(s+n)}{n^s n!}$$
$$\times \frac{(1-s)(2-s)\cdots(n+1-s)}{n^{1-s}n!}$$
$$= s\lim_{n\to\infty}\left(1 + \frac{1-s}{n}\right)\prod_{k=1}^{n}\left(1 - \frac{s^2}{k^2}\right),$$

which is equal to

$$s\prod_{n=1}^{\infty}\left(1 - \frac{s^2}{n^2}\right) = \frac{\sin\pi s}{\pi}$$

by **PROBLEM 2.11**. $\square$

REMARK. This formula can be used to evaluate Raabe's integral (**PROBLEM 16.7**). For, it follows that

$$\int_0^1 \log\Gamma(s)\,ds = \frac{1}{2}\left(\int_0^1 \log\Gamma(s)\,ds + \int_0^1 \log\Gamma(1-s)\,ds\right)$$
$$= \frac{1}{2}\int_0^1 \log\frac{\pi}{\sin\pi s}\,ds.$$

It is not hard to see that the last expression is equal to $\log\sqrt{2\pi}$, which was the value of $c = f(1) + 1$.

SOLUTION 16.9

Let $\phi(s)$ denote the integral on the right-hand side of the equality to be shown. For any fixed closed subinterval $[a, b]$ of $(0, \infty)$ we have

$$\left(s - 1 - \frac{1 - e^{-(s-1)x}}{1 - e^{-x}}\right)\frac{e^{-x}}{x} = \frac{(s-1)(s-3)}{2} + O(x)$$

as $x \to 0+$, where the constant in O-symbol is uniform in $s \in [a, b]$. Similarly we have

$$\left(s - 1 - \frac{1 - e^{-(s-1)x}}{1 - e^{-x}}\right)\frac{e^{-x}}{x} = \begin{cases} (s-2)\dfrac{e^{-x}}{x} + o(e^{-\delta x}) & \text{for} \quad s > 1, \\[2mm] 0 & \text{for} \quad s = 1, \\[2mm] \dfrac{e^{-sx}}{x} + o(e^{-x}) & \text{for} \quad 0 < s < 1, \end{cases}$$

where $\delta = \max\{2, s\}$ and the constant in the estimates is uniform in $s \in [a, b]$. This implies that $\phi \in C(0, \infty)$.

For any positive numbers s and t,

$$\frac{\phi(s) + \phi(t)}{2} = \int_0^\infty \left(\frac{s+t}{2} - 1 - \frac{1 - (e^{-sx} + e^{-tx})e^x/2}{1 - e^{-x}}\right)\frac{e^{-x}}{x}\,dx$$

$$\geq \phi\left(\frac{s+t}{2}\right),$$

since e^{-sx} is convex with respect to s and $1 - e^{-x} > 0$ for any $x > 0$. Therefore ϕ is convex on the interval $(0, \infty)$.

Since $\phi(1) = 0$ is obvious, it follows from the Bohr-Mollerup theorem that $\phi(s) = \log\Gamma(s)$ if only we show that $\phi(s+1) - \phi(s) = \log s$ for any $s > 0$. However it is easily verified that

$$\phi(s+1) - \phi(s) = \int_0^\infty \frac{e^{-x} - e^{-sx}}{x}\,dx,$$

which can be seen to be $\log s$ by integrating

$$\int_0^\infty e^{-sx}\,dx = \frac{1}{s}$$

with respect to s.   □

SOLUTION 16.10

It follows from Malmstén's formula that

$$\log \Gamma(s+y) = \int_0^\infty \left(s + y - 1 - \frac{1 - e^{-(s+y-1)x}}{1 - e^{-x}} \right) \frac{e^{-x}}{x} \, dx$$

for any $0 \le y \le 1$. Integrating with respect to y we get

$$\int_s^{s+1} \log \Gamma(x) \, dx = \int_0^\infty \left(s - \frac{1}{2} + \frac{e^{-(s-1)x}}{x} - \frac{1}{1 - e^{-x}} \right) \frac{e^{-x}}{x} \, dx,$$

whose left-hand side is, by Raabe's formula (**PROBLEM 16.7**), equal to

$$s \log s - s + \frac{1}{2} \log(2\pi).$$

Hence, subtracting

$$\frac{1}{2} \log s = \frac{1}{2} \int_0^\infty \frac{e^{-x} - e^{-sx}}{x} \, dx$$

from this, we see that

$$\left(s - \frac{1}{2} \right) \log s - s + \frac{1}{2} \log(2\pi) =$$

$$\int_0^\infty \left(s - 1 + \left(\frac{1}{2} + \frac{1}{x} \right) e^{-(s-1)x} - \frac{1}{1 - e^{-x}} \right) \frac{e^{-x}}{x} \, dx.$$

By Malmstén's formula again, we conclude that

$$\omega(s) = \int_0^\infty \left(-\frac{1}{2} - \frac{1}{x} + \frac{1}{1 - e^{-x}} \right) \frac{e^{-sx}}{x} \, dx,$$

as asserted.   □

SOLUTION 16.11

Since the function $\log \Gamma(s)$ is asymptotic to $- \log s$ as $s \to 0+$, the improper integral

$$\int_0^1 \log \frac{\Gamma(s)}{\Gamma(1 - s)} \, ds$$

converges absolutely; hence it follows from **PROBLEM 7.11** that, for any $0 < s < 1$,

$$\log \frac{\Gamma(s)}{\Gamma(1 - s)} = \frac{a_0}{2} + \sum_{n=1}^\infty \left(a_n \cos 2n\pi x + b_n \sin 2n\pi x \right)$$

where a_n and b_n are the Fourier coefficients defined in the remark after
PROBLEM 7.11.

By the change of variable $\sigma = 1 - s$ we have

$$a_n = 2 \int_0^1 \log \frac{\Gamma(1 - \sigma)}{\Gamma(\sigma)} \cos 2n\pi\sigma \, d\sigma = -a_n \, ;$$

hence $a_n = 0$ for all $n \geq 0$.

To calculate b_n we use the formula in **PROBLEM 16.1**:

$$\log \frac{\Gamma(s)}{\Gamma(1 - s)} = \log \frac{1 - s}{s} + \lim_{m \to \infty} A_m(s),$$

where

$$A_m(s) = (2s - 1) \log m + \sum_{k=1}^m \log \frac{k + 1 - s}{k + s}.$$

We show that the limit on the right-hand side converges uniformly in $s \in [0, 1]$.
To this end, note that, for any positive integer $p > q$, we have

$$\left| \sum_{k=q+1}^p \frac{1}{k + s} - \log \frac{p}{q} \right| < \frac{1}{q + 1}$$

for any $0 \leq s \leq 1$; therefore

$$\left| A_p(s) - A_q(s) \right| = \left| (2s - 1) \log \frac{p}{q} + \sum_{k=q+1}^p \log \left(1 + \frac{1 - 2s}{k + s} \right) \right|$$

$$\leq \sum_{k=q+1}^p \left| \log \left(1 + \frac{1 - 2s}{k + s} \right) + \frac{2s - 1}{k + s} \right| + \frac{1}{q + 1}$$

$$< \frac{2}{q},$$

where we used the inequality $| \log(1 + x) - x| \leq x^2$. Thus we can interchange the
order of limit and integration to obtain

$$b_n = 2 \lim_{m \to \infty} \int_0^1 \widetilde{A}_m(s) \sin 2n\pi s \, ds,$$

where

$$\widetilde{A}_m(s) = (2s - 1) \log m + \sum_{k=0}^m \log \frac{k + 1 - s}{k + s}.$$

Problems and Solutions in Real Analysis

Since

$$\int_0^1 (2s - 1) \sin 2n\pi s \, ds = -\frac{1}{n\pi}$$

and

$$\sum_{k=0}^{m-1} \int_0^1 \log \frac{k + 1 - s}{k + s} \sin 2n\pi s \, ds = -2 \int_0^m \log t \sin 2n\pi t \, dt,$$

we obtain

$$b_n = -2 \lim_{m \to \infty} \left(\frac{\log m}{n\pi} + 2 \int_0^m \log t \sin 2n\pi t \, dt \right).$$

Here we ignored the last term corresponding to $k = m$ in $\widetilde{A}_m(s)$, since it converges to 0 uniformly in s. It follows by integration by parts that

$$\int_0^m \log t \sin 2n\pi t \, dt = \left[\frac{1 - \cos 2n\pi t}{2n\pi} \log t \right]_{t=0+}^{t=m}$$

$$- \frac{1}{2n\pi} \int_0^m \frac{1 - \cos 2n\pi t}{t} \, dt$$

$$= -\frac{1}{2n\pi} \int_0^1 \frac{1 - \cos 2mn\pi x}{x} \, dx.$$

Hence

$$b_n = \frac{2(\log n + C)}{n\pi},$$

where

$$C = \lim_{N \to \infty} \left(\int_0^1 \frac{1 - \cos 2N\pi x}{x} \, dx - \log N \right)$$

is a constant independent of n. Then it is easily seen that

$$C = \lim_{N \to \infty} \left(\int_0^{2N\pi} \frac{1 - \cos s}{s} \, ds - \log N \right)$$

$$= \int_0^1 \frac{1 - \cos s}{s} \, ds - \int_1^\infty \frac{\cos s}{s} \, ds + \log 2\pi,$$

which is equal to $\gamma + \log 2\pi$ by **Problem 6.9**. We thus have

$$\log \frac{\Gamma(s)}{\Gamma(1 - s)} = \frac{2}{\pi} \sum_{n=1}^\infty \frac{\log n + \gamma + \log 2\pi}{n} \sin 2n\pi x,$$

which implies Kummer's series by virtue of **PROBLEM 7.10** and **PROBLEM 16.8**. □

REMARK. Kummer's series converges uniformly on each interval $[\delta, 1 - \delta]$ for any $\delta > 0$ by Dirichlet's test.

Chapter 17

Prime Number Theorem

Let $\pi(x)$ be the number of primes not exceeding x. The prime number theorem states that

$$\lim_{x \to \infty} \frac{\pi(x) \log x}{x} = 1,$$

in other words, $\pi(x)$ is asymptotic to $x/\log x$ as $x \to \infty$, which is one of the most celebrated results in mathematics. The first major step toward the prime number theorem was made by Chebyshev (1852), who showed that

$$A \frac{x}{\log x} \le \pi(x) \le A' \frac{x}{\log x}$$

for sufficiently large x where

$$A = \log \left(\frac{2^{1/2} 3^{1/3} 5^{1/5}}{30^{1/30}} \right) = 0.92129... \quad \text{and} \quad A' = \frac{6}{5} A = 1.10555...$$

- Chebyshev introduced therein two important functions

$$\theta(x) = \sum_{p \le x} \log p \quad \text{and} \quad \psi(x) = \sum_{p^k \le x} \log p$$

with a real variable x, where p runs over primes and k over positive integers. Note that

$$\psi(n) = \log [1, 2, ..., n]$$

where $[1, 2, ..., n]$ denotes the least common multiple of $1, 2, ..., n$.

- The Chebyshev function $\psi(x)$ can also be written as

$$\psi(x) = \sum_{p \le x} \Lambda(p),$$

239

where

$$\Lambda(n) = \begin{cases} \log p & \text{if } n \text{ is a power of a prime } p, \\ 0 & \text{otherwise}, \end{cases}$$

known as the von Mangoldt function, introduced in 1895. We define $\Lambda(x) = \Lambda([x])$ for a real variable $x \geq 1$ for convenience.

- The Möbius function $\mu(n)$, introduced in Möbius (1832), is defined by $\mu(n) = (-1)^k$ if n is a product of k distinct primes (for $n = 1$ we set $\mu(1) = 1$) and by $\mu(n) = 0$ otherwise. It is easily seen that

$$\sum_{d|n} \mu(d) = \begin{cases} 1 & \text{if } n = 1, \\ 0 & \text{if } n \geq 2, \end{cases}$$

where d runs over all of the divisors of n.

- Let f and g be any functions defined on all of the divisors of a positive integer n. The Möbius inversion formula states that if $g(n) = \sum_{d|n} f(d)$, then

$$f(n) = \sum_{d|n} \mu(d) g\left(\frac{n}{d}\right).$$

As an application of this formula we have

$$\Lambda(n) = -\sum_{d|n} \mu(d) \log d,$$

as the inversion of

$$\sum_{d|n} \Lambda(d) = \sum_{p^k|n} \log p = \sum k \log p = \sum \prod p^k = \log n.$$

The prime number theorem was first established by Hadamard (1896) and by de la Vallée Poussin (1896) independently, following Riemann's program in 1859. Their arguments are based on the nonvanishing of $\zeta(z)$ on the vertical line $\Re z = 1$ and also on the existence of some zero-free region of $\zeta(z)$ in the critical strip $0 \leq \Re z \leq 1$. Wiener's Tauberian theory on Fourier analysis implies the equivalence of the prime number theorem and the nonvanishment of $\zeta(z)$ on $\Re z = 1$. Later Newman (1980) found a simple analytic proof of the prime number theorem. See also Korevaar (1982) and Zagier (1997).

Selberg (1949) and Erdös (1949) succeeded in giving elementary proofs of the prime number theorem, in the sense that they did not use the Riemann zeta function, complex analysis, or Fourier analysis. Note that "elementary" does not necessarily mean "simple".

For some fine historical perspective of the prime number theorem, see Levinson (1969), Goldstein (1973), Diamond (1982) and Bateman and Diamond (1996).

Below we shall give an elementary proof of the prime number theorem in the spirit of Erdös and Selberg, culminating in **PROBLEM 17.11**.

PROBLEM 17.1

Show that

$$\sum_{n \le x} \psi\left(\frac{x}{n}\right) = \log [x]!$$

for any $x \ge 1$, where n runs over positive integers not exceeding x.

This is due to Chebyshev (1852).

PROBLEM 17.2

Let $f(x)$ be any function defined for $x \ge 1$ and put

$$g(x) = (\log x) \sum_{n \le x} f\left(\frac{x}{n}\right).$$

Show then that

$$\sum_{n \le x} \mu(n) g\left(\frac{x}{n}\right) = f(x) \log x + \sum_{n \le x} \Lambda(n) f\left(\frac{x}{n}\right).$$

This is due to Tatuzawa and Iseki (1951).

PROBLEM 17.3

Apply the formula stated in **PROBLEM 17.2** *for $f(x) = \psi(x) - x + \gamma + 1$ to deduce that*

$$\psi(x) \log x + \sum_{n \le x} \Lambda(n) \psi\left(\frac{x}{n}\right) = 2x \log x + O(x)$$

as $x \to \infty$, where γ is Euler's constant.

This is also due to Tatuzawa and Iseki (1951), which can be shown to be equivalent to

$$\theta(x)\log x + \sum_{p \leq x} \theta\left(\frac{x}{p}\right)\log p = 2x\log x + O(x),$$

as $x \to \infty$, known as Selberg's inequality (1949) where p runs over primes not exceeding x. Note that the sum on the left-hand side is non-negative, which implies immediately that $\psi(x) = O(x)$ as $x \to \infty$. This can be usually derived from the consideration of a binomial coefficient. See the remark after SOLUTION 17.3.

PROBLEM 17.4

Show that

$$\pi(x) = \frac{\theta(x)}{\log x} + \int_2^x \frac{\theta(t)}{t\log^2 t}\,dt$$

for any $x \geq 2$. Deduce from this that the prime number theorem is equivalent to $\psi(x) \sim x$ as $x \to \infty$.

It is easily seen that the prime number theorem is also equivalent to $\theta(x) \sim x$ as $x \to \infty$.

PROBLEM 17.5

Show that

$$U(x)\log x + \int_1^x \Lambda(t)U\left(\frac{x}{t}\right)dt = O(x),$$

as $x \to \infty$, where

$$U(x) = \int_1^x \frac{\psi(t) - t}{t}\,dt.$$

Levinson (1966) introduced $U(x)$ defined above, which is a Lipschitz function since $\psi(x) = O(x)$, to show the prime number theorem by proving that $U(x) = o(x)$ as $x \to \infty$. To see this, for any small $\epsilon > 0$, let x_ϵ be a positive number satisfying

$$(1 - \epsilon)x < \int_1^x \frac{\psi(t)}{t}\,dt < (1 + \epsilon)x$$

for any $x > x_\epsilon$. Putting $y = \left(1 + \sqrt{\epsilon}\,\right)x$ we obtain

$$(y - x)\frac{\psi(x)}{y} \le \int_x^y \frac{\psi(t)}{t}\,dt = \int_0^y \frac{\psi(t)}{t}\,dt - \int_0^x \frac{\psi(t)}{t}\,dt$$
$$< (1 + \epsilon)y - (1 - \epsilon)x;$$

therefore

$$\frac{\psi(x)}{x} < \left(1 + \sqrt{\epsilon}\,\right)^3.$$

By a similar way we have

$$\frac{1 - 2\sqrt{\epsilon} - \epsilon}{1 + \sqrt{\epsilon}} < \frac{\psi(y)}{y},$$

which implies that $\psi(x) \sim x$ as $x \to \infty$; namely, the prime number theorem by **Problem 17.4**. Therefore the remainder is devoted to show that $U(x) = o(x)$. Our method is substantially based on the works of Wright (1954) and Levinson (1969), however we replace finite sums by integrals as often as possible.

Problem 17.6 ───────────────────────────────────

Iterate the estimate given in **Problem 17.5** *to show that*

$$U(x)\log^2 x + \iint_{\Delta_x} (\Lambda(st) - \Lambda(s)\Lambda(t)) U\left(\frac{x}{st}\right)ds\,dt = O(x\log x)$$

as $x \to \infty$, *where* Δ_x *is the region defined by* $st \le x$ *and* $s, t \ge 1$.

Problem 17.7 ───────────────────────────────────

Let $x > 1$, f *be continuous and* g *be a Lipschitz function with constant* L *satisfying* $g(1) = 0$ *defined on the interval* $[1, x]$. *Show then that*

$$\left|\int_1^x f(t)g\left(\frac{x}{t}\right)dt\right| \le Lx \int_1^x \left|\int_1^t f(s)ds\right|\frac{dt}{t^2}.$$

If $g(x)$ is continuously differentiable on the interval $[1, x]$, then $|g'(t)| \le L$ and it follows by integration by parts that

$$\int_1^x f(t)g\left(\frac{x}{t}\right)dt = \left[F(t)g\left(\frac{x}{t}\right)\right]_{t=1}^{t=x} + x\int_1^x F(t)g'\left(\frac{x}{t}\right)\frac{dt}{t^2},$$

which implies immediately the above inequality, where $F(t) = \int_1^t f(s)ds$. This observation would be a good hint.

PROBLEM 17.8

Show that

$$\iint_{\Delta_x} (\Lambda(st) + \Lambda(s)\Lambda(t))\,ds\,dt = 2x\log x + O(x)$$

as $x \to \infty$. Then using **PROBLEMS 17.6** and **17.7** prove that

$$|U(x)|\log^2 x \le 2\iint_{\Delta_x}\left|U\left(\frac{x}{st}\right)\right|ds\,dt + Cx\log x \qquad (17.1)$$

for some constant C.

PROBLEM 17.9

Putting $V(x) = e^{-x}U(e^x)$ show that

$$\limsup_{x\to\infty}|V(x)| = \limsup_{x\to\infty}\frac{1}{x}\int_0^x |V(s)|\,ds.$$

PROBLEM 17.10

Let $f(x)$ be a bounded Lipschitz function with a constant L defined on the interval $[0, \infty)$. Put

$$\alpha = \limsup_{x\to\infty}|f(x)| \quad and \quad \beta = \limsup_{x\to\infty}\frac{1}{x}\int_0^x |f(s)|\,ds.$$

Suppose further that

$$\delta = \limsup_{x\to\infty}\left|\int_0^x f(s)\,ds\right|$$

is finite. Show then that

$$\beta(\alpha^2 + 2\delta L) \le 2\alpha\delta L.$$

PROBLEM 17.11

Deduce the prime number theorem from the above facts.

Solutions for Chapter 17

| SOLUTION 17.1 |

It follows from the definition of ψ that

$$\sum_{n \le x} \psi\left(\frac{x}{n}\right) = \sum_{mn \le x} \Lambda(m) = S.$$

In the last sum m and n run over all positive integers satisfying $mn \le x$. We now rearrange such pairs (m, n) according to the value of $k = mn$; namely,

$$S = \sum_{k \le x} \sum_{mn = k} \Lambda(m).$$

In the second sum m runs over all of the divisors of k. Hence we have

$$S = \sum_{k \le x} \sum_{d \mid k} \Lambda(d) = \sum_{k \le x} \log k = \log[x]!.$$

$\square$

| SOLUTION 17.2 |

By definition,

$$\sum_{n \le x} \mu(n) g\left(\frac{x}{n}\right) = \sum_{mn \le x} \mu(n) f\left(\frac{x}{mn}\right) \log \frac{x}{n}.$$

In the last sum we rearrange pairs (m, n) in the same manner as in SOLUTION 17.1; then we have

$$= \sum_{k \le x} f\left(\frac{x}{k}\right) \sum_{d \mid k} \mu(d) \log \frac{x}{d}$$

$$= (\log x) \sum_{k \le x} f\left(\frac{x}{k}\right) \sum_{d \mid k} \mu(d) - \sum_{k \le x} f\left(\frac{x}{k}\right) \sum_{d \mid k} \mu(d) \log d.$$

This implies the desired formula, since the first term is $f(x) \log x$ and in the second $\sum_{d \mid k} \mu(d) \log d = -\Lambda(k)$, as is shown in the introduction. $\square$

| SOLUTION 17.3 |

It follows from **PROBLEM 17.1** and from Stirling's approximation that

$$S = \sum_{n \le x} \psi\left(\frac{x}{n}\right) = \log [x]!$$
$$= [x] \log [x] - [x] + O(\log x)$$
$$= x \log x - x + O(\log x)$$

as $x \to \infty$. We next have

$$I = \int_1^\infty \frac{\{x\}}{x^2} \, dx = \sum_{k=1}^{n-1} \int_k^{k+1} \frac{x-k}{x^2} \, dx + O\left(\frac{1}{n}\right)$$
$$= \log n - \sum_{k=2}^n \frac{1}{k} + O\left(\frac{1}{n}\right)$$

as $n \to \infty$, where $\{x\}$ denotes the fractional part of x. This implies $I = 1 - \gamma$ and therefore

$$\sum_{n \le x} \frac{1}{n} = \log x + \gamma + O\left(\frac{1}{x}\right)$$

as $x \to \infty$. Hence, for $f(x) = \psi(x) - x + \gamma + 1$ specified in the problem, it follows that

$$\sum_{n \le x} f\left(\frac{x}{n}\right) = x \log x - x - x \log x - \gamma x + (\gamma + 1)x + O(\log x)$$
$$= O(\log x).$$

Hence, for $g(x)$ stated in **PROBLEM 17.2**, we may choose a positive constant K satisfying $|g(x)| \le K \log^2 x$ and we have

$$\left| \sum_{n \le x} \mu(n) g\left(\frac{x}{n}\right) \right| \le K \sum_{n \le x} \log^2\left(\frac{x}{n}\right)$$
$$\le K \log^2 x + K \int_1^x \log^2\left(\frac{x}{t}\right) dt$$
$$< K \log^2 x + Kx \int_1^\infty \left(\frac{\log t}{t}\right)^2 dt = O(x)$$

as $x \to \infty$.   Substituting this estimate and the expression for $f(x)$ in **PROBLEM 17.2**, thereby putting

$$\phi(x) = \psi(x) \log x + \sum_{n \le x} \Lambda(n) \psi\left(\frac{x}{n}\right), \tag{17.2}$$

we obtain

$$\phi(x) = x \log x + x \sum_{n \leq x} \frac{\Lambda(n)}{n} + O(x) + O(\psi(x)).$$

On the other hand, for the sum S in **Solution 17.1**, we have

$$S = \sum_{mn \leq x} \Lambda(n) = \sum_{n \leq x} \left[\frac{x}{n}\right] \Lambda(n) = x \sum_{n \leq x} \frac{\Lambda(n)}{n} + O(\psi(x)),$$

and that

$$\phi(x) = 2x \log x + O(x) + O(\psi(x)).$$

This also implies that $\psi(x) \log x = 2x \log x + O(x) + O(\psi(x))$, since the sum in (17.2) is non-negative. Therefore $\psi(x) = O(x)$ and so $\phi(x) = 2x \log x + O(x)$, as required.   □

Remark. The fact $\psi(x) = O(x)$ is usually shown, as follows. The binomial coefficient

$$\binom{2m+1}{m} = \frac{(2m+1)2m \cdots (m+1)}{m!}$$

is a multiple of the product of all primes lying in the interval $(m, 2m+1]$, which is clearly less than $(1+1)^{2m+1}/2 = 2^{2m}$, since it appears twice in the binomial expansion of $(1+1)^{2m+1}$. Therefore

$$\theta(2m+1) - \theta(m) = \sum_{m < p \leq 2m+1} \log p < 2m \log 2$$

holds for any positive integer m. Hence for any $x \geq 2$, putting $k = [x/2]$, we obtain

$$\theta(x) - \theta\left(\frac{x}{2}\right) \leq \theta(2k+2) - \theta(k)$$

$$= \theta(2k+1) - \theta(k) < 2k \log 2 \leq x \log 2;$$

so,

$$\theta(x) = \sum_{n \geq 0} \left(\theta\left(\frac{x}{2^n}\right) - \theta\left(\frac{x}{2^{n+1}}\right)\right) < 2x \log 2.$$

We thus have

$$\psi(x) = \theta(x) + \theta\left(x^{1/2}\right) + \theta\left(x^{1/3}\right) + \cdots + \theta\left(x^{1/N}\right)$$
$$< (2 \log 2)\left(x + x^{1/2} + x^{1/3} + \cdots + x^{1/N}\right)$$
$$< 2x \log 2 + 2\sqrt{x} \log x = O(x),$$

where $N = [\log x / \log 2]$.

SOLUTION 17.4

Let m be any positive integer. Since the difference $\theta(m) - \theta(m - 1)$ is equal to $\log m$ if n is a prime and to 0 otherwise, we have by partial summation

$$\pi(n) = \sum_{m=2}^{n} \frac{\theta(m) - \theta(m - 1)}{\log m}$$
$$= \frac{\theta(n)}{\log n} + \sum_{m=2}^{n-1} \theta(m)\left(\frac{1}{\log m} - \frac{1}{\log(m + 1)}\right).$$

By using

$$\frac{1}{\log m} - \frac{1}{\log(m + 1)} = \int_{m}^{m+1} \frac{dt}{t \log^2 t},$$

we get

$$\pi(n) = \frac{\theta(n)}{\log n} + \int_{2}^{n} \frac{\theta(t)}{t \log^2 t} \, dt,$$

on noting that $\theta(t) = \theta(m)$ in the interval $[m, m + 1)$. Moreover we can replace n by a real variable $x \geq 2$ in view of $\pi(x) = \pi([x])$ and

$$\int_{[x]}^{x} \frac{\theta(t)}{t \log^2 t} \, dt = \theta([x])\left(\frac{1}{\log [x]} - \frac{1}{\log x}\right).$$

Since there exists a positive constant C satisfying $\theta(x) \leq Cx$ for any $x \geq 2$ by the remark after SOLUTION 17.3,

$$\int_{2}^{x} \frac{\theta(t)}{t \log^2 t} \, dt \leq C \int_{2}^{x} \frac{dt}{\log^2 t}.$$

Integrating by parts we get

$$\int_2^x \frac{dt}{\log^2 t} = \left[\frac{t}{\log^2 t} \right]_{t=2}^{t=x} + 2 \int_2^x \frac{dt}{\log^3 t}$$

$$\leq \frac{x}{\log^2 x} + C' + \frac{1}{2} \int_2^x \frac{dt}{\log^2 t}$$

for some constant C', since $2/\log^3 t \leq 1/(2\log^2 t)$ for any $t \geq e^4$. Therefore

$$\int_2^x \frac{\theta(t)}{t \log^2 t} dt = O\left(\frac{x}{\log^2 x} \right)$$

and

$$\frac{\pi(x)\log x}{x} - \frac{\theta(x)}{x} = O\left(\frac{1}{\log x} \right)$$

as $x \to \infty$, which implies that the prime number theorem is equivalent to $\theta(x) \sim x$. This completes the proof, since

$$\psi(x) - \theta(x) = O(\sqrt{x}\log x)$$

as is already seen in the remark after **SOLUTION 17.3**. □

SOLUTION 17.5

We have already shown in **SOLUTION 17.3** that

$$\sum_{n \leq x} \frac{\Lambda(n)}{n} = \log x + O(1)$$

as $x \to \infty$, which implies immediately that for $R(x) = \psi(x) - x$ for $x \geq 1$,

$$R(x)\log x + \sum_{n \leq x} \Lambda(n) R\left(\frac{x}{n} \right) = O(x).$$

Replacing x by t, dividing by t and integrating from 1 to x with respect to t we have

$$\int_1^x \frac{R(t)}{t}\log t \, dt + \int_1^x \sum_{n \leq t} \Lambda(n) R\left(\frac{t}{n} \right) \frac{dt}{t} = O(x). \qquad (17.3)$$

Integrating by parts, the first integral on the left-hand side of (17.3) becomes

$$U(x)\log x - \int_1^x \frac{U(t)}{t} dt = U(x)\log x + O(x),$$

Problems and Solutions in Real Analysis

since $U(x) = O(x)$. On the other hand, the second integral is, with $\widetilde{R}(x) = 0$ for $0 \leq x < 1$, $\widetilde{R}(x) = R(x)$ for $x \geq 1$ and $\widetilde{U}(x) = \int_0^x \widetilde{R}(s)/s\,ds$,

$$\int_1^x \sum_{n=1}^{\infty} \Lambda(n)\widetilde{R}\left(\frac{t}{n}\right)\frac{dt}{t} = \sum_{n=1}^{\infty} \Lambda(n)\int_{1/n}^{x/n} \frac{\widetilde{R}(s)}{s}\,ds$$

$$= \sum_{n=1}^{\infty} \Lambda(n)\widetilde{U}\left(\frac{x}{n}\right);$$

therefore the second integral differs from $\int_1^x \Lambda(t)\,U(x/t)\,dt$ by

$$\sum_{n=1}^{\infty} \Lambda(n)\int_n^{n+1} \left(\widetilde{U}\left(\frac{x}{n}\right) - \widetilde{U}\left(\frac{x}{t}\right)\right)dt,$$

which is clearly of order $O(x)$ since $\widetilde{U}(x)$ is a Lipschitz function. Note that the sums are actually finite.   □

SOLUTION 17.6

Replacing x by x/s in the estimate given in **PROBLEM 17.5**, multiplying by $\Lambda(s)$ and integrating from $s = 1$ to $s = x$ we get

$$(\log x)\int_1^x \Lambda(s)\,U\left(\frac{x}{s}\right)ds - \int_1^x \Lambda(s)(\log s)\,U\left(\frac{x}{s}\right)ds$$

$$+ \iint_{\Delta_x} \Lambda(s)\Lambda(t)\,U\left(\frac{x}{st}\right)ds\,dt = O(x\log x),$$

since we have

$$\int_1^x \frac{\Lambda(s)}{s}\,ds = \sum_{n\leq x} \frac{\Lambda(n)}{n} + O(1)$$

and this is $\log x + O(1)$ as $x \to \infty$, as proved in **SOLUTION 17.3**. Since the second integral on the left-hand side is

$$\int_1^x \Lambda(s)(\log s)\,U\left(\frac{x}{s}\right)ds = \iint_{\Delta_x} \Lambda(st)\,U\left(\frac{x}{st}\right)ds\,dt,$$

the estimate in question follows immediately from **PROBLEM 17.5**.   □

SOLUTION 17.7

By the uniform continuity of g, for any $\epsilon > 0$, there exists a $\delta \in (0, 1)$ such that $|g(s) - g(t)| < \epsilon$ whenever $s, t \in [1, x]$ satisfy $|s - t| < \delta$. Instead of the Lipschitz function $g(t)$ we now consider, for $t \geq 1$,

$$G(t) = \frac{1}{\delta} \int_{t-\delta}^{t} \tilde{g}(s)\,ds,$$

where $\tilde{g}(s) = 0$ for $0 < s < 1$ and $\tilde{g}(s) = g(s)$ for $s \geq 1$. $G(t)$ is continuously differentiable and satisfies $G(1) = 0$. Since $\tilde{g}$ is also a Lipschitz function with constant L, we have

$$|G'(t)| = \frac{|\tilde{g}(t) - \tilde{g}(t - \delta)|}{\delta} \leq L.$$

Moreover, since $G(t) = \tilde{g}(\xi_t)$ for some $\xi_t \in (t - \delta, t)$, G is very close to g in the sense that

$$|G(t) - g(t)| = |\tilde{g}(\xi_t) - g(t)| < \epsilon$$

for any $t \in [1, x]$. Hence

$$\left| \int_1^x f(t) g\left(\frac{x}{t}\right) dt \right| \leq \epsilon \int_1^x |f(t)|\,dt + \left| \int_1^x f(t) G\left(\frac{x}{t}\right) dt \right|.$$

By integration by parts, we obtain

$$\int_1^x f(t) G\left(\frac{x}{t}\right) dt = \left[\left(\int_1^t f(s)\,ds \right) G\left(\frac{x}{t}\right) \right]_{t=1}^{t=x}$$
$$+ x \int_1^x \left(\int_1^t f(s)\,ds \right) G'\left(\frac{x}{t}\right) \frac{dt}{t^2}$$

and therefore, by the inequality for $G'(t)$ above,

$$\left| \int_1^x f(t) g\left(\frac{x}{t}\right) dt \right| \leq \epsilon \int_1^x |f(t)|\,dt + Lx \int_1^x \left| \int_1^t f(s)\,ds \right| \frac{dt}{t^2}.$$

This completes the proof, since ϵ is arbitrary.   □

SOLUTION 17.8

Since, by partial integration,

$$\int_1^x \Lambda(v) \log \frac{x}{v}\,dv = \int_1^x \left(\int_1^v \Lambda(u)\,du \right) \frac{dv}{v} = O(x)$$

    Problems and Solutions in Real Analysis

as $x \to \infty$, it follows that

$$\iint_{\Delta_x} \Lambda(st)\,ds\,dt = \int_1^x \Lambda(v) \log v\,dv = (\log x) \int_1^x \Lambda(v)\,dv + O(x)$$

$$= \psi(x) \log x + O(x).$$

We also have, by the above proven result,

$$\iint_{\Delta_x} \Lambda(s)\Lambda(t)\,ds\,dt = \int_1^x \Lambda(t) \int_1^{x/t} \Lambda(s)\,ds\,dt$$

$$= \int_1^x \Lambda(t)\left(\psi\left(\frac{x}{t}\right) + O\left(\log\frac{x}{t}\right)\right)dt$$

$$= \int_1^x \Lambda(t)\psi\left(\frac{x}{t}\right)dt + O(x)$$

$$= \sum_{n \le x} \Lambda(n)\psi\left(\frac{x}{n}\right) + O(x);$$

thus the first estimate in the problem follows from **PROBLEM 17.3**.

The second estimate follows from **PROBLEM 17.6** if we show that

$$\iint_{\Delta_x} (\Lambda(st) + \Lambda(s)\Lambda(t) - 2)\left|U\left(\frac{x}{st}\right)\right| ds\,dt = O(x\log x). \tag{17.4}$$

To see this observe that the left-hand side can be written as

$$\int_1^x \Phi(v)\left|U\left(\frac{x}{v}\right)\right| dv$$

where

$$\Phi(v) = \int_1^v \left(\Lambda(v) + \Lambda(u)\Lambda\left(\frac{v}{u}\right) - 2\right)\frac{du}{u}.$$

Since

$$\int_1^y \Phi(v)\,dv = \iint_{\Delta_y} (\Lambda(st) + \Lambda(s)\Lambda(t) - 2)\,ds\,dt = O(y)$$

and $|U(t)|$ is a Lipschitz function, (17.4) follows from **PROBLEM 17.7**.   □

| **SOLUTION 17.9** |

Writing $x = e^u$ and making the change of variables $x/s = e^v$, $x/(st) = e^w$ in the integral on the right-hand side of (17.1), the region corresponding to Δ_x is

$\{(v, w); 0 \leq w \leq v \leq u\}$ and the Jacobian is $e^{u-w} > 0$. Hence, putting $V(x) = e^{-x}U(e^x)$ and dividing by $u^2 e^u$, (17.1) reads

$$|V(u)| \leq \frac{2}{u^2} \int_0^u \int_0^v |V(w)| \, dw \, dv + \frac{C}{u}.$$

For brevity put

$$\alpha = \limsup_{u \to \infty} |V(u)| \quad \text{and} \quad \beta = \limsup_{v \to \infty} \frac{1}{v} \int_0^v |V(w)| \, dw.$$

By definition, for any $\epsilon > 0$, there exists a $v_\epsilon > 0$ such that

$$\int_0^v |V(w)| \, dw < (\beta + \epsilon)v$$

for any $v > v_\epsilon$. We then have, using the above estimate,

$$\alpha \leq \limsup_{u \to \infty} \frac{2}{u^2} \left(C_\epsilon u + \frac{\beta + \epsilon}{2} u^2 \right) = \beta + \epsilon$$

where

$$C_\epsilon = \int_0^{v_\epsilon} |V(w)| \, dw,$$

whence $\alpha \leq \beta$. This completes the proof, the reverse inequality $\alpha \geq \beta$ being obvious.   □

SOLUTION 17.10

We can assume that $\beta > 0$. Then the function $f(x)$ must have arbitrarily large zeros. For otherwise, the sign of $f(x)$ does not change on the interval $[x_0, \infty)$ for some x_0 and we can find a divergent sequence $x_0 < x_1 < x_2 < \cdots$ satisfying

$$\int_0^{x_n} |f(s)| \, ds > \frac{\beta}{2} x_n$$

for all $n \geq 1$. Therefore

$$\left| \int_0^{x_n} f(s) \, ds \right| \geq \int_{x_1}^{x_n} |f(s)| \, ds - \int_0^{x_1} |f(s)| \, ds$$

$$\geq \frac{\beta}{2} x_n - 2 \int_0^{x_1} |f(s)| \, ds,$$

which diverges to ∞, contrary to the assumption that δ is finite.

It follows from the definition of superior limits that, for any $\epsilon > 0$ there is a point $x_\epsilon > 0$ satisfying

$$|f(x)| \le \alpha + \epsilon \quad \text{and} \quad \left| \int_0^x f(s)\,ds \right| < \delta + \epsilon$$

for any $x \ge x_\epsilon$. For brevity put

$$\kappa_\epsilon = \frac{8(\delta + \epsilon)L}{(\alpha + \epsilon)^2 + 2(\delta + \epsilon)L} \quad \text{and} \quad \mu_\epsilon = 1 + \frac{2(\delta + \epsilon)L}{(\alpha + \epsilon)^2}.$$

Note that μ_ϵ is always $\ge \kappa_\epsilon$.

Let now $a < b$ be any two consecutive zero points $\ge x_\epsilon$ of $f(x)$. We distinguish three cases as follows, according to the value of $\sigma = (b - a)L/(\alpha + \epsilon)$.

Case (i) $\sigma < \kappa_\epsilon$. Since the Lipschitz condition implies that the graph of $|f(x)|$ on the interval $[a, b]$ is contained in the equilateral triangular region with base $[a, b]$ and in height $(b - a)L/2$, we have clearly

$$\frac{1}{b - a} \int_a^b |f(s)|\,ds \le \frac{(b - a)L}{4} < \frac{\kappa_\epsilon(\alpha + \epsilon)}{4}.$$

Case (ii) $\kappa_\epsilon \le \sigma \le \mu_\epsilon$. By the same reason the graph of $|f(x)|$ is contained in the same triangular region as above. However in this case, since $|f(x)| \le \alpha + \epsilon$, we can replace it by the smaller trapezoidal region in height $\alpha + \epsilon$ with the same base. Hence

$$\frac{1}{b - a} \int_a^b |f(s)|\,ds \le \frac{(\alpha + \epsilon)^2}{(b - a)L} + (\alpha + \epsilon)\left(1 - \frac{2(\alpha + \epsilon)}{(b - a)L}\right)$$

$$\le (\alpha + \epsilon)\left(1 - \frac{1}{\mu_\epsilon}\right).$$

Case (iii) $\sigma > \mu_\epsilon$. Since

$$\int_a^b |f(s)|\,ds = \left| \int_a^b f(s)\,ds \right| \le \left| \int_0^a f(s)\,ds \right| + \left| \int_0^b f(s)\,ds \right|$$

$$< 2(\delta + \epsilon),$$

we have

$$\frac{1}{b - a} \int_a^b |f(s)|\,ds < \frac{2(\delta + \epsilon)}{b - a} < \frac{2(\delta + \epsilon)L}{\mu_\epsilon(\alpha + \epsilon)}.$$

The upper bounds derived in each case coincide with each other and therefore

$$\frac{1}{b-a} \int_a^b |f(s)| ds < \frac{2(\alpha + \epsilon)(\delta + \epsilon)L}{(\alpha + \epsilon)^2 + 2(\delta + \epsilon)L}.$$

For any $x > x_\epsilon$ let a^* and b^* be the smallest and the largest zero points of $f(x)$ in the interval $[x_\epsilon, x]$ respectively. Since the sign of $f(x)$ on the interval $(b^*, x]$ does not change, it follows that

$$\int_0^x |f(s)| ds \le \int_0^{a^*} |f(s)| ds + \sum_{a < b} \int_a^b |f(s)| ds + 2(\delta + \epsilon);$$

hence

$$\beta \le \frac{2(\alpha + \epsilon)(\delta + \epsilon)L}{(\alpha + \epsilon)^2 + 2(\delta + \epsilon)L}.$$

Letting $\epsilon \to 0+$ we get the desired inequality.   □

SOLUTION 17.11

Suppose first that the function $V(x)$ defined in **PROBLEM 17.9** satisfies all of the conditions stated in **PROBLEM 17.10**. We then would have $\alpha(\alpha^2 + 2\delta L) \le 2\alpha\delta L$, which implies $\alpha = 0$; namely, $U(x) = o(x)$ as $x \to \infty$. Therefore for the proof of the prime number theorem it suffices to show that (i) $V(x)$ is a Lipschitz function and (ii) $\int_0^x V(s) ds$ is bounded.

Let $0 \le x < y$ be any numbers. We have

$$
\begin{aligned}
|V(y) - V(x)| &= \left| e^{-y} U(e^y) - e^{-x} U(e^x) \right| \\
&\le e^{-y} |U(e^y) - U(e^x)| + (e^{-x} - e^{-y}) |U(e^x)| \\
&\le e^{-y} \int_{e^x}^{e^y} \frac{|\psi(t) - t|}{t} dt + (e^{-x} - e^{-y}) |U(e^x)| \\
&\le C(1 - e^{x-y})
\end{aligned}
$$

for some constant C, since the integrand is a bounded function. Thus $V(x)$ is a Lipschitz function in view of $1 - e^{x-y} < y - x$.

As is already seen, the integral

$$\int_1^x \frac{\Lambda(s) - 1}{s} ds$$

Problems and Solutions in Real Analysis

is bounded for $x \geq 1$, which is equal to

$$\frac{1}{x} \int_1^x (\Lambda(s) - 1)\, ds + \int_1^x \frac{1}{s^2} \int_1^s (\Lambda(t) - 1)\, dt\, ds$$

by integration by parts. Since

$$\int_1^s (\Lambda(t) - 1)\, dt = \psi(s) - s + O(\log s),$$

the integral

$$\int_1^x \frac{\psi(s) - s}{s^2}\, ds$$

is also bounded. Again it follows by integration by parts that

$$\frac{U(x)}{x} + \int_1^x \frac{U(s)}{s^2}\, ds$$

is bounded, and therefore we get the boundedness of the integral

$$\int_1^x \frac{U(s)}{s^2}\, ds = \int_0^{\log x} V(s)\, ds,$$

as required. This completes the proof of the prime number theorem. $\square$

Chapter 18

Miscellanies

- The nth *Bernoulli number* B_n is defined as the coefficient of z^{2n} in the Taylor series of the function $z/(e^z - 1)$ about $z = 0$:

$$\frac{z}{e^z - 1} = 1 - \frac{z}{2} + \sum_{n=1}^{\infty} (-1)^{n-1} \frac{B_n}{(2n)!} z^{2n}$$

with the radius of convergence 2π.

The Bernoulli numbers were introduced by Jakob Bernoulli (1654–1705) in the book 'Ars Conjectandi' published in 1713, eight years after his death. They have a connection with the evaluation of sum of the powers of positive integers. Each B_n is a positive rational number. The first nine B_n's are:

$$B_1 = \frac{1}{6}, \quad B_2 = \frac{1}{30}, \quad B_3 = \frac{1}{42}, \quad B_4 = \frac{1}{30}, \quad B_5 = \frac{5}{66},$$

$$B_6 = \frac{691}{2730}, \quad B_7 = \frac{7}{6}, \quad B_8 = \frac{3617}{510}, \quad B_9 = \frac{43867}{798}.$$

The reader should notice that the numbering and the signs of the Bernoulli numbers may be different in some books.

PROBLEM 18.1

Suppose that $a(x) \in C(\mathbb{R})$ satisfies $m \le a(x) \le M$ for two constants m and M. Show that the first-order differential equation

$$y' = y(a(x) - y)$$

has a solution $y(x)$ satisfying $m \le y(x) \le M$ for any real x.

PROBLEM 18.2

Suppose that $a(x)$ and $b(x)$ are continuous functions on the interval $[0, \infty)$ satisfying $a(x) > 0$ for all $x \ge 0$,

$$\int_0^\infty a(x)\, dx = \infty \quad and \quad K = \int_0^\infty \frac{b^2(x)}{a(x)}\, dx < \infty.$$

Prove that a solution of the first-order linear differential equation

$$y' = a(x)y + b(x)$$

satisfying

$$\int_0^\infty a(x)y^2(x)\, dx < \infty \tag{18.1}$$

is uniquely determined and satisfies

$$\int_0^\infty a(x)y^2(x)\, dx \le K. \tag{18.2}$$

PROBLEM 18.3

Show that any solution $y(x)$ of the second-order linear differential equation for $x > 0$

$$y'' + xy = 0$$

is bounded and has an infinite number of zero points.

PROBLEM 18.4

Prove the irrationality of π^2 by expressing the integral

$$I_n = \int_0^\pi P(x) \sin x \, dx$$

as a linear combination of $1, \pi^2, \pi^4, ..., \pi^{2n}$ *over* $\mathbb{Z}$, *where*

$$P(x) = \frac{1}{(2n)!} x^{2n} (\pi - x)^{2n}.$$

This is due to Niven (1947), giving a very simple irrationality proof of π.

PROBLEM 18.5

Show that the Bernoulli number B_n satisfies

$$B_n \equiv (-1)^n \left(\frac{1}{2} + \sum_d{}^* \frac{1}{2d+1} \right) \quad (\text{mod } 1)$$

for any positive integer n where d runs over all of the divisors of n such that $2d + 1$ is a prime.

For example,

$$1 - B_1 = \frac{1}{2} + \frac{1}{3}, \quad 1 + B_2 = \frac{1}{2} + \frac{1}{3} + \frac{1}{5}, \quad 1 - B_3 = \frac{1}{2} + \frac{1}{3} + \frac{1}{7},$$

$$1 + B_4 = \frac{1}{2} + \frac{1}{3} + \frac{1}{5}, \quad 1 - B_5 = \frac{1}{2} + \frac{1}{3} + \frac{1}{11},$$

$$1 + B_6 = \frac{1}{2} + \frac{1}{3} + \frac{1}{5} + \frac{1}{7} + \frac{1}{13}, \quad 2 - B_7 = \frac{1}{2} + \frac{1}{3}.$$

This is known as the Staudt-Clausen theorem, which is found by Staudt (1840) and by Clausen (1840) independently. The latter was published in 'Astronomische Nachrichten', the oldest astronomical journal founded in 1821 by H. C. Schumacher (one of the friends and astronomical collaborators of Gauss), as a brief announcement of the result without proof. Later Schwering (1899) gave another proof using

$$\frac{1}{x+1} + \frac{1}{2(x+1)(x+2)} + \cdots + \frac{(n-1)!}{n(x+1)(x+2)\cdots(x+n)} + \cdots$$

$$= \frac{1}{x} - \frac{1}{2x^2} + \frac{B_1}{x^3} - \frac{B_2}{x^5} + \frac{B_3}{x^7} - \cdots$$

and Wilson's theorem, which gave rise to the proof by Kluyver (1900) using Fermat's little theorem.

Clausen was a lover of numbers. He calculated correctly the 248 decimals of π in 1847 (Delahaye (1997)) and factored the 6th Fermat number $2^{64} + 1$ to

$$274, 177 \times 67, 280, 421, 310, 721$$

in 1854 (Schönbeck (2004)).

PROBLEM 18.6

Generalize the method employed in **SOLUTION 10.3** *to obtain*

$$\zeta(2n) = \frac{2^{2n-1}}{(2n)!} B_n \pi^{2n}. \qquad (18.3)$$

This is due to Apostol (1973). Williams (1971) gave a similar proof but used complex function theory in addition. Skau and Selmer (1971) also discussed a similar method to obtain the rationality of $\zeta(2n)/\pi^{2n}$. See also Hovstad (1972). Chen (1975) gave a proof on the same lines. Robbins (1999) gave another proof for Chen's recursive formula, using the Fourier expansion of x^{2n} on the interval $[-\pi, \pi]$. Ji and Chen (2000) proved inductively using the Fourier expansion of some quadratic function and iterated integrations.

On the other hand, Kuo (1949) obtained a rather complicated formula which represents $\zeta(2n)$ as a sum of $\zeta(2)$, ..., $\zeta(2[n/2])$ and their products, while the proof requires the Fourier series appeared in **PROBLEM 7.11** and Parseval's theorem. Kuo's formula is equivalent to a formula for the Bernoulli numbers, for which Carlitz (1961) gave a simpler and slightly general formula using the Bernoulli polynomials.

Stark (1972) considered the even moments of the Dirichlet kernel

$$D_n(\theta) = \frac{\sin(n + 1/2)\theta}{2\sin(\theta/2)} = \frac{1}{2} + \sum_{k=1}^{n} \cos k\theta,$$

as a generalization of his earlier work in 1969, in which he had used the Fejér kernel. See the comment after **PROBLEM 10.4**. He also pointed out that the method employed in **SOLUTION 10.7** is implicitly based upon the de la Vallée Poussin kernel. Again Stark (1974) used the Fejér kernel to give a simpler proof the recursive formula obtained in 1972.

Berndt (1975) gave two elementary proofs, the first is based on the calculation of the Fourier coefficients of the Bernoulli polynomials, and the second is based on the partial fraction expansion

$$\pi^2 \operatorname{cosec}^2 \pi x = \sum_{k=-\infty}^{\infty} \frac{1}{(k + x)^2},$$

whose elementary proof was given by Neville (1951).

Osler (2004) used the product formula of sines presented in **PROBLEM 2.11**. Tsumura (2004) evaluated $\zeta(2m)$ in a different way.

PROBLEM 18.7

Show that

$$\left(n + \frac{1}{2}\right)\zeta(2n) = \sum_{k=1}^{n-1} \zeta(2k)\zeta(2n - 2k)$$

for any positive integer $n \geq 2$ without using the Bernoulli numbers.

This is due to Williams (1953), which enables us to determine the value of $\zeta(2n)$ from $\zeta(2)$. Two somewhat complicated recursive formulae satisfied by $\zeta(2k)$ were already given in Titchmarsh (1926). Estermann (1947) gave a recursive formula

$$(2^{2n} - 1)\zeta(2n) = \sum_{k=1}^{n-1} a_{k,n}\,\zeta(2k)\zeta(2n - 2k)$$

where $a_{1,n} = 2^{2n} - 10$ and $a_{k,n} = -2(2k-1)(2^{2n-2k} - 1)$ for $k \geq 2$. Their formulae are obtained only by rearranging absolutely convergent series.

In order to deduce (18.3) from this recursive formula, it suffices to show that

$$(2n + 1)B_n = \sum_{k=1}^{n-1}\binom{2n}{2k}B_k B_{n-k}$$

for any integer $n \geq 2$, which was proved by Underwood (1928). The direct proof from the definition of the Bernoulli numbers can be found in Berndt (1975), as follows.

Let

$$f(z) = \frac{z}{e^z - 1} = 1 - \frac{z}{2} + g(z)$$

for brevity. By definition, the coefficient of z^{2n} for $n \geq 2$ in the Taylor expansion of $(z f(z))'$ about $z = 0$ is

$$(-1)^{n-1}\frac{2n + 1}{(2n)!}B_n.$$

On the other hand,

$$(z f(z))' = (2 - z)f(z) - f^2(z) = \left(1 - \frac{z}{2}\right)^2 - g^2(z)$$

and the coefficient of z^{2n} for $n \geq 2$ on the right-hand side is equal to

$$-(-1)^n \sum_{k=1}^{n-1} \frac{B_k B_{n-k}}{(2k)!(2n-2k)!},$$

which implies the desired recursive formula.

Solutions for Chapter 18

SOLUTION 18.1

Note that the right-hand side is a quadratic function in y; hence it does not satisfy the Lipschitz condition. This means that a local solution may blow up at a finite x. For example, in the specific case $a(x) = m = M = 0$ every general solution with constant C

$$y(x) = \frac{1}{x + C}$$

blows up at $x = -C$. In this case the only solution satisfying the condition of problem is the trivial solution: $y = 0$.

Without loss of generality we can assume that the interval $[m, M]$ does not contain the point 0; for otherwise the trivial solution $y = 0$ would satisfy the required condition. We can moreover assume that $M \geq m > 0$; for otherwise consider $\tilde{y}(x) = -y(-x)$ and $\tilde{a}(x) = -a(x)$. By the substitution $y = 1/w$ we get a first-order linear differential equation of the form

$$w' = 1 - a(x)w \tag{18.4}$$

so that y is a global solution of the original equation if and only if $w = 1/y$ is a global solution of (18.4) with constant sign. Putting

$$A(x) = \int_0^x a(s)\,ds,$$

the general solution of (18.4) with constant C_0 can be written as

$$w(x) = e^{-A(x)}\left(\int_0^x e^{A(s)}\,ds + C_0 \right).$$

Since $A(x)$ diverges to $-\infty$ as $x \to -\infty$, the factor $e^{-A(x)}$ diverges to ∞. This means that we must take

$$C_0 = -\int_0^{-\infty} e^{A(s)}\,ds;$$

otherwise $w(x)$ would be unbounded; hence,

$$w(x) = e^{-A(x)} \int_{-\infty}^{x} e^{A(s)} ds = \int_{-\infty}^{x} \exp\left(-\int_{s}^{x} a(t)\,dt\right) ds.$$

Using the inequality $a(x) \geq m > 0$ we get

$$w(x) \leq \int_{-\infty}^{x} e^{-m(x-s)} ds = \int_{0}^{\infty} e^{-mt}\,dt = \frac{1}{m}.$$

We obtain $w(x) \geq 1/M$ similarly, which means that $y = 1/w$ satisfies $m \leq y \leq M$.

$\square$

REMARK. The first-order differential equation of the form

$$y' + p(x)y = q(x)y^{n}$$

is called the Bernoulli differential equation. By the substitution $w = y^{1-n}$ if $n \neq 1$ it reduces to a first-order linear differential equation.

SOLUTION 18.2

Putting

$$A(x) = \int_{0}^{x} a(s)\,ds$$

for brevity, the general solution with constant C can be written in the form

$$y(x) = e^{A(x)}\left(\int_{0}^{x} b(s)e^{-A(s)}\,ds + C\right).$$

By the assumption $A(x)$ diverges to ∞ as $x \to \infty$, which means that we must take

$$C = -\int_{0}^{\infty} b(s)e^{-A(s)}\,ds;$$

otherwise $y(x)$ would be unbounded and would not satisfy the inequality (18.1). Note that, by the Cauchy-Schwarz inequality,

$$\left(\int_{0}^{\infty} |b(s)|e^{-A(s)}\,ds\right)^{2} \leq \int_{0}^{\infty} \frac{b^{2}(s)}{a(s)}\,ds \cdot \int_{0}^{\infty} a(s)e^{-2A(s)}\,ds = \frac{K}{2}$$

in view of

$$\int_{0}^{\infty} a(s)e^{-2A(s)}\,ds = \left[-\frac{1}{2}e^{-2A(s)}\right]_{s=0}^{s=\infty} = \frac{1}{2}$$

and so this improper integral converges absolutely.

We next show that the solution chosen above

$$y(x) = -e^{A(x)} \int_x^\infty b(s)e^{-A(s)} ds$$

satisfies the inequality (18.2). In the same way as above we have

$$y^2(x) \le e^{2A(x)} \int_x^\infty \frac{b^2(s)}{a(s)} ds \cdot \int_x^\infty a(s)e^{-2A(s)} ds$$

$$= \frac{1}{2} \int_x^\infty \frac{b^2(s)}{a(s)} ds,$$

which converges to 0 as $x \to \infty$. By integrating the equation $ay^2 = yy' - by$ from 0 to L, we obtain

$$\sigma_L = \int_0^L a(x)y^2(x) dx = \frac{y^2(L) - y^2(0)}{2} - \int_0^L b(x)y(x) dx$$

$$\le \frac{y^2(L)}{2} + \int_0^L |b(x)y(x)| dx.$$

By a similar argument as above we have

$$\left(\int_0^L |b(x)y(x)| dx \right)^2 \le \sigma_L \int_0^L \frac{b^2(x)}{a(x)} dx \le K\sigma_L.$$

Putting $\epsilon_L = y^2(L)/2$ we thus get $\sigma_L \le \epsilon_L + \sqrt{K\sigma_L}$. Let σ^* be a unique positive solution of the equation $\sigma = \epsilon_L + \sqrt{K\sigma}$. Then

$$\sigma_L \le \sigma^* = \epsilon_L + \frac{K}{2} + \frac{\sqrt{K^2 + 4K\epsilon_L}}{2}$$

$$\le \epsilon_L + \frac{K}{2} + \frac{K + 2\epsilon_L}{2} = K + 2\epsilon_L.$$

Since ϵ_L converges to 0 as $L \to \infty$, we obtain

$$\int_0^\infty a(x)y^2(x) dx \le K,$$

as required.   □

SOLUTION 18.3

We will show that any solution $y(x)$ has at least one zero point in the open interval $J_k = (k\pi, (k+1)\pi)$ for every positive integer k. Suppose, on the contrary, that $y(x) \ne 0$ for any $x \in J_k$ for some k. Putting

$$\varphi(x) = y(x)\cos x - y'(x)\sin x,$$

we have

$$\varphi'(x) = y'(x)\cos x - y(x)\sin x - y''(x)\sin x - y'(x)\cos x$$
$$= (x-1)y(x)\sin x.$$

Hence the function φ is monotone on J_k, since $\mathrm{sgn}(\varphi') = (-1)^k \mathrm{sgn}(y)$.
On the other hand, let $\Delta\varphi$ be the difference $\varphi((k+1)\pi) - \varphi(k\pi)$. Since

$$(-1)^{k+1}\Delta\varphi = y(k\pi) + y((k+1)\pi)$$

and $\mathrm{sgn}(\Delta\varphi) = \mathrm{sgn}(\varphi')$ on J_k, we get

$$\begin{aligned}
\mathrm{sgn}(y) &= (-1)^k \mathrm{sgn}(\varphi') \\
&= (-1)^k \mathrm{sgn}(\Delta\varphi) \\
&= -\mathrm{sgn}\Big(y(k\pi) + y((k+1)\pi)\Big),
\end{aligned}$$

a contradiction.

We next show that any solution $y(x)$ satisfies the inequality

$$y^2(x) \le y^2(1) + (y'(1))^2$$

for any $x > 1$. To see this, multiplying the differential equation by y' and integrating from 1 to x we have

$$\int_1^x y'(t)y''(t)\,dt + \int_1^x ty(t)y'(t)\,dt = 0.$$

By applying integration by parts to the second integral, it follows that

$$\Big[(y'(t))^2\Big]_{t=1}^{t=x} + \Big[ty^2(t)\Big]_{t=1}^{t=x} = \int_1^x y^2(t)\,dt;$$

therefore

$$xu(x) \le c + \int_1^x u(t)\,dt$$

where $u(x) = y^2(x)$ and $c = y^2(1) + (y'(1))^2$; namely, if we put

$$v(x) = \frac{1}{x}\int_1^x u(t)\,dt,$$

we get

$$v'(x) = \frac{u(x)}{x} - \frac{1}{x^2}\int_1^x u(t)\,dt \le \frac{c}{x^2}.$$

Integrating from 1 to x again,

$$v(x) \le c \int_1^x \frac{dt}{t^2} = c\left(1 - \frac{1}{x}\right),$$

which implies $\int_1^x u(t)\,dt \le c(x-1)$; hence $u(x) \le c$.   □

REMARK. Putting

$$y(x) = \sum_{n=0}^{\infty} a_n x^n,$$

one can easily obtain a power series solution by solving the recursive formula

$$a_n = \begin{cases} 0 & \text{for } n = 2, \\ -\dfrac{a_{n-3}}{n(n-1)} & \text{for } n \ge 3. \end{cases}$$

Hence we obtain

$$a_n = \begin{cases} \dfrac{(-1)^{n/3} a_0}{n(n-1)(n-3)(n-4)\cdots 3\cdot 2} & \text{for } n \equiv 0 \pmod 3, \\ \dfrac{(-1)^{[n/3]} a_1}{n(n-1)(n-3)(n-4)\cdots 4\cdot 3} & \text{for } n \equiv 1 \pmod 3, \\ 0 & \text{for } n \equiv 2 \pmod 3, \end{cases}$$

which implies that the radius of convergence is ∞; in other words, every solution is real analytic.

SOLUTION 18.4

Integrating by parts repeatedly we have

$$\int_0^\pi P(x) e^{ix}\,dx = -i \left[P(x) e^{ix}\right]_{x=0}^{x=\pi} + i \int_0^\pi P'(x) e^{ix}\,dx$$

$$\vdots$$

$$= i(P(0) + P(\pi)) + i^3(P'(0) + P''(\pi))$$
$$+ \cdots + i^{4n+1}(P^{(4n)}(0) + P^{(4n)}(\pi)).$$

Since $P^{(k)}(0) = P^{(k)}(\pi) = 0$ for any $0 \le k < 2n$ and $P(x) = P(\pi - x)$, we obtain

$$I_n = \sum_{k=n}^{2n} (-1)^k \left(P^{(2k)}(0) + P^{(2k)}(\pi) \right)$$

$$= 2 \sum_{k=n}^{2n} (-1)^k P^{(2k)}(0).$$

On the other hand, if $n \le k \le 2n$,

$$P^{(2k)}(x) = \frac{1}{(2n)!} \sum_{j=0}^{2n} (-1)^j \binom{2n}{j} \pi^{2n-j} (x^{2n+j})^{(2k)}$$

$$= \sum_{j=2k-2n}^{2n} (-1)^j \frac{(2n+j)!}{j!(2n-j)!(2n+j-2k)!} \pi^{2n-j} x^{2n+j-2k},$$

which implies that

$$P^{(2k)}(0) = \frac{(2k)!}{(2k-2n)!(4n-2k)!} \pi^{4n-2k};$$

therefore

$$I_n = 2 \sum_{\ell=0}^{n} (-1)^\ell \frac{(4n-2\ell)!}{(2\ell)!(2n-2\ell)!} \pi^{2\ell}.$$

Suppose now that $\pi^2 = p/q$ for some positive integers p and q. Since $I_n > 0$, $q^n I_n$ is a positive integer and less than

$$\frac{q^n}{(2n)!} \int_0^\pi x^{2n}(\pi - x)^{2n}\, dx < \frac{q^n \pi^{4n}}{2^{4n}(2n)!}.$$

Obviously the right-hand side converges to 0 as $n \to \infty$, a contradiction.   □

REMARK. The formula used in the above proof can be obtained by substituting $z = \pi i$ in the $(2n, 2n)$-Padé approximation to e^z. One can prove the irrationality of the numbers $\log \alpha$ with $\alpha \in \mathbb{Q}$ and $\alpha \ne 1$ in the same way.

SOLUTION 18.5

This is based on Kluyver's proof. In a neighborhood of $z = 0$ we have $|e^z - 1| < 1$; hence, expanding $z = \log(1 + e^z - 1)$ in $e^z - 1$ and dividing by $e^z - 1$, we get

$$f(z) = \frac{z}{e^z - 1} = 1 - \frac{e^z - 1}{2} + \frac{(e^z - 1)^2}{3} - \cdots.$$

Hence

$$B_n = (-1)^{n-1} f^{(2n)}(0)$$

$$= (-1)^n \left(\frac{1}{2} - \frac{c_{2,2n}}{3} + \frac{c_{3,2n}}{4} - \cdots - \frac{c_{2n,2n}}{2n+1} \right),$$

where

$$c_{k,m} = \frac{d^m}{dz^m} (e^z - 1)^k \bigg|_{z=0}.$$

The basic fact here is that if $\phi(z) = (e^z - 1)^m$, then $\phi^{(k)}(0)$ is always divisible by $m!$. Therefore, if $k + 1 = ab$ is a composite number greater than 4, then $a + b \le k$ and

$$c_{k,2n} = \frac{d^{2n}}{dz^{2n}} \left((e^z - 1)^a (e^z - 1)^b (e^z - 1)^{k-a-b} \right) \bigg|_{z=0}$$

is divisible by $a!b!(k-a-b)!$; thus $c_{k,2n}/(k+1)$ is an integer. Moreover if $k = 3$, then

$$c_{3,2n} = 3 - 3 \cdot 2^{2n} + 3^{2n} \equiv 0 \pmod 4$$

so that $c_{3,2n}/4$ is an integer.

If $k + 1 = p$ is an odd prime, then, since $\ell^k \equiv 1 \pmod p$ by Fermat's little theorem, it follows from

$$c_{k,m} = \sum_{\ell=1}^{k} (-1)^{k-\ell} \binom{k}{\ell} \ell^m$$

that $c_{k,m} \pmod p$ is periodic in m with period k. Furthermore we know that $c_{k,1} = c_{k,2} = \cdots = c_{k,k-1} = 0$ and

$$c_{k,k} \equiv \sum_{\ell=1}^{k} (-1)^{k-\ell} \binom{k}{\ell} \equiv -1 \pmod p.$$

Hence

$$c_{p-1,2n} \equiv -1 \pmod p$$

if and only if $2n$ is divisible by $p - 1$; otherwise

$$c_{p-1,2n} \equiv 0 \pmod p.$$

$\square$

REMARK. It can be easily seen that $c_{k,k} = k!$ for any positive integer k. Thus it follows from Wilson's theorem that $c_{k,k} \equiv -1 \pmod{p}$.

SOLUTION 18.6

Since $\cot^2\theta < \theta^{-2} < 1 + \cot^2\theta$ for $0 < \theta < \pi/2$, we have

$$\cot^{2n}\theta < \frac{1}{\theta^{2n}} < (1 + \cot^2\theta)^n.$$

As is already seen in **SOLUTION 10.3**, the m roots of the polynomial

$$\varphi(x) = \sum_{k=0}^{m} (-1)^k \binom{2m+1}{2k+1} x^{m-k}$$

of degree m are

$$x_k = \cot^2\frac{k\pi}{2m+1}$$

for $1 \leq k \leq m$. Putting $s_n = x_1^n + \cdots + x_m^n$ and using the inequalities mentioned above we get

$$s_n < \frac{(2m+1)^{2n}}{\pi^{2n}} \sum_{k=1}^{m} \frac{1}{k^{2n}} < \sum_{k=1}^{m} (1 + x_k)^n,$$

where the right-hand side is expanded as

$$s_n + \binom{n}{1} s_{n-1} + \cdots . \tag{18.5}$$

Thus if s_n is asymptotic to $c_{2n} m^{2n}$ for some constant c_{2n} as $m \to \infty$, then (18.5) is asymptotic to the same one and we conclude that

$$\zeta(2n) = c_{2n} \left(\frac{\pi}{2}\right)^{2n}.$$

Since s_n is a symmetric function, it can be expressed as a sum of elementary symmetric functions. Indeed, by Newton's formula we have

$$\binom{2m+1}{1} s_n - \binom{2m+1}{3} s_{n-1} + \cdots$$
$$+ (-1)^{n-1} \binom{2m+1}{2n-1} s_1 + (-1)^n \binom{2m+1}{2n+1} n = 0$$

when $n \leq m$. Using this formula we will show

$$s_n \sim \frac{2^{4n-1}}{(2n)!} B_n m^{2n}$$

by induction on n. Obviously this holds true for $n = 1$ in view of $c_2 = 2/3 = 4B_1$. Suppose now that this holds true up to n. Then by Newton's formula

$$\lim_{m \to \infty} \frac{S_{n+1}}{m^{2n+2}}$$

exists and is equal to

$$\frac{2^2}{3!} c_{2n} - \frac{2^4}{5!} c_{2n-2} + \cdots - (-1)^n \frac{2^{2n}}{(2n+1)!} c_2 + (-1)^n \frac{(n+1)2^{2n+2}}{(2n+3)!}.$$

By induction assumption this is written as

$$(-1)^{n-1} 2^{4n+4} \left(\sum_{k=1}^{n} (-1)^{k-1} \frac{2^{-(2n-2k+3)}}{(2n-2k+3)!} \cdot \frac{B_k}{(2k)!} - \frac{n+1}{2^{2n+2}(2n+3)!} \right).$$

The expression in the parentheses added to

$$\frac{(-1)^n}{2(2n+2)!} B_{n+1}$$

is exactly equal to the coefficient of z^{2n+3} in the expansion of

$$\frac{z e^{z/2}}{e^z - 1}.$$

However this function is even and the coefficient of z^{2n+3} vanishes. Hence we obtain

$$\lim_{m \to \infty} \frac{S_{n+1}}{m^{2n+2}} = \frac{2^{4n+3}}{(2n+2)!} B_{n+1},$$

which completes the induction. $\qquad\qquad\square$

SOLUTION 18.7

We start with

$$\sum_{k=1}^{n-1} \zeta(2k)\zeta(2n-2k) = \lim_{m \to \infty} S_m,$$

where

$$S_m = \sum_{k=1}^{n-1} \sum_{j=1}^{m} \sum_{\ell=1}^{m} \frac{1}{j^{2k} \ell^{2n-2k}}$$

$$= (n-1) \sum_{j=1}^{m} \frac{1}{j^{2n}} + \sum_{1 \le j \ne \ell \le m} \frac{\ell^{-2n+2} - j^{-2n+2}}{j^2 - \ell^2}.$$

Let T_m denote the second sum on the right-hand side. Since

$$\sum_{1 \le j \ne \ell \le m} \frac{-j^{-2n+2}}{j^2 - \ell^2} = \sum_{1 \le j \ne \ell \le m} \frac{-\ell^{-2n+2}}{\ell^2 - j^2},$$

it follows that

$$T_m = 2 \sum_{1 \le j \ne \ell \le m} \frac{1}{\ell^{2n-2}(j^2 - \ell^2)} = 2 \sum_{\ell=1}^{m} \frac{c_\ell}{\ell^{2n-2}},$$

where

$$c_\ell = \sum \frac{1}{j^2 - \ell^2}$$

and the summation runs through the integral value of j in $[1, m]$ except for the value ℓ. We then have

$$c_\ell = \frac{1}{2\ell} \sum \left(\frac{1}{j - \ell} - \frac{1}{j + \ell} \right) = \frac{3}{4\ell^2} - \frac{1}{2\ell} \sum_{j=m-\ell+1}^{m+\ell} \frac{1}{j},$$

so that

$$T_m = \frac{3}{2} \sum_{\ell=1}^{m} \frac{1}{\ell^{2n}} + O\left(\sum_{\ell=1}^{m} \frac{1}{\ell^{2n-1}} \sum_{j=m-\ell+1}^{m+\ell} \frac{1}{j} \right).$$

Note that the cancellation is similar to that in **Solution 10.1**. Since

$$\sum_{j=m-\ell+1}^{m+\ell} \frac{1}{j} < \frac{2\ell}{m - \ell + 1},$$

we get, for the error term for T_m,

$$\sum_{\ell=1}^{m} \frac{2\ell}{\ell^{2n-1}(m - \ell + 1)} \le 2 \sum_{\ell=1}^{m} \frac{1}{\ell(m - \ell + 1)}$$

$$= \frac{2}{m+1} \sum_{\ell=1}^{m} \left(\frac{1}{\ell} + \frac{1}{m - \ell + 1} \right)$$

$$= O\left(\frac{\log m}{m} \right)$$

as $m \to \infty$. Therefore

$$S_m = \left(n + \frac{1}{2} \right) \sum_{j=1}^{m} \frac{1}{j^{2n}} + O\left(\frac{\log m}{m} \right),$$

which completes the proof.   □

Bibliography

[**Books**]

Achieser, N. I. (1956). Theory of Approximation, Frederick Ungar Pub. Co., New York.
[Translated by C. J. Hyman from the author's lectures on approximation theory at Univ. Kharkov.]

Ahlfors, L. V. (1966). Complex Analysis, 2nd edition, McGraw-Hill.

Apostol, T. M. (1957). Mathematical Analysis – A Modern Approach to Advanced Calculus, Addison-Wesley.

Artin, E. (1964). The Gamma Function, Holt, Rinehart and Winston, New York.
[Translated by M. Butler from 'Einführung in die Theorie der Gammafunktion', Hamburger Math. Einzelschriften, Verlag B. G. Teubner, Leipzig, 1931.]

Bass, J. (1966). Exercises in Mathematics : Simple and multiple integrals. Series of functions. Fourier series and Fourier integrals. Analytic functions. Ordinary and partial differential equations., Academic Press, New York-London.
[Translated by Scripta Technica, Inc. from 'Exercices de Mathématiques. Algèbre linéaire. Intégrales simples et multiples. Séries de fonctions. Séries et intégrales de Fourier. Fonctions analytiques. Equations différentielles et aux dérivées partielles. Calcul des probabilités', Masson et Cie, Éditeurs, Paris, 1965.]

Berndt, B. C. (1994). Ramanujan's Notebooks, Part IV, Springer-Verlag.

Bernoulli, Johannis (1697). Opera Omnia, Tom I, Edited by J. E. Hofmann, Georg Olms Verlagsbuchhandlung, Hildesheim, 1968 (in particular, pp. 184–185).

Borwein, J. M. and Borwein, P. B. (1987). Pi and the AGM, Wiley Interscience Pub. (in particular, p. 381).

Cauchy (1841). Exercices d'analyse et de physique mathématique, Vol. 2, Bachelier, Paris (in particular, p. 380).

Cheney, E. W. (1966). Introduction to Approximation Theory, McGraw-Hill Book Co., New York.

Delahaye, J.-P. (1997). Le fascinant nombre π, Pour la Science, Diffusion Belin, Paris (in particular, p. 60).

Dieudonné, J. (1971). Infinitesimal Calculus, Hermann, Paris.
[Translated from 'Calcul infinitésimal', 1968, Hermann, Paris.]

Finch, S. R. (2003). Mathematical Constants, Encyclopedia of Math. and its Appl. 94, Cambridge Univ. Press.

274   *Problems and Solutions in Real Analysis*

Hardy, G. H. (1958). A Course of Pure Mathematics, Cambridge Univ. Press.
————— (1963). Divergent Series, 3rd ed., Oxford Univ. Press.
Hardy, G. H., Littlewood, J. E. and Pólya, G. (1934). Inequalities, Cambridge Univ. Press.
Hardy, G. H. and Wright, E. M. (1979). An Introduction to the Theory of Numbers, 5th ed., Oxford Univ. Press.
Hobson, E. W. (1957). The Theory of Functions of a Real Variable and the Theory of Fourier's Series, Vol. II, Dover edition.
Kac, M. (1972). Statistical Independence in Probability, Analysis and Number Theory, The Carus Math. Monographs, John Wiley & Sons.
Kanemitsu, S. and Tsukada, H. (2007). Vistas of Special Functions, World Scientific Publishing Company.
Korevaar, J. (2004). Tauberian Theory – A Century of Developments, Springer-Verlag, Berlin.
Korobov, N. M. (1992). Exponential Sums and their Applications, Kluwer Acad. Pub.
    [Translated from 'Trigonometrical Sums and their Applications', Nauka, Moscow, 1989.]
Lang, S. (1965). Algebra, Addison-Wesley.
le Lionnais, F. (1983). Les nombres remarquables, Hermann, Paris (in particular, p. 22).
Levin, L. (1981). Polylogarithms and Associated Functions, North-Holland, New York (in particular, p. 4).
Niven, I. and Zuckerman, H. S. (1960). An Introduction to the Theory of Numbers, New York, John Wiley & Sons.
Pólya, G. and Szegö, G. (1972). Problems and Theorems in Analysis I, Springer-Verlag New York Berlin Heidelberg.
    [Translated from 'Aufgaben und Lehrsätze aus der Analysis I', 4th ed., 1970, Heiderberger Taschenbücher, Bd. 73.]
————————— (1976). Problems and Theorems in Analysis II, Springer-Verlag New York Berlin Heidelberg.
    [Translated from 'Aufgaben und Lehrsätze aus der Analysis II', 4th ed., 1971, Heiderberger Taschenbücher, Bd. 74.]
Stirling, J. (1730). Methodus Differentialis : sive tractatus de summatione et interpolatione serierum infinitarum, G. Bowyer, G. Strahan, London.
Wall, H. S. (1948). Analytic Theory of Continued Fractions, D. Van Nostrand Company, Inc., Princeton (in particular, p. 331).
Zygmund, A. (1979). Trigonometric Series, Vol. 1, Cambridge Univ. Press.

[Papers]

Abel, N. H. (1826). Recherches sur la série $1 + \frac{m}{1}x + \frac{m(m-1)}{1 \cdot 2}x^2 + \frac{m(m-1)(m-2)}{1 \cdot 2 \cdot 3}x^3 + \cdots$, *J. Reine Angew. Math.* **1**, pp. 311–339.
    = Œuvres complètes de Niels Henrik Abel, Nouvelle Édition, Tome I, Christiania Grøndahl & Søn, 1881, pp. 219–250 (in particular, p. 223).
Apéry, R. (1979). Irrationalité de $\zeta(2)$ et $\zeta(3)$, *Astérisque* **61**, pp. 11–13.

Apostol, T. M. (1973). Another elementary proof of Euler's formula for $\zeta(2n)$, *Amer. Math. Monthly* **80**, no. 4, pp. 425–431.

———————— (1983). A proof that Euler missed: Evaluating $\zeta(2)$ the easy way, *Math. Intelligencer* **5**, pp. 59–60.

Arratia, A. (1999). Algunas maneras juveniles de evaluar $\zeta(2k)$, *Bol. Asoc. Mat. Venez.* **6**, no. 2, pp. 167–176.

Ayoub, R. (1974). Euler and the zeta function, *Amer. Math. Monthly* **81**, no. 10, pp. 1067–1086.

Barnes, E. W. (1899). The theory of the Gamma function, *Mess. Math.* **29**, pp. 64–128.

Bateman, P. T. and Diamond, H. G. (1996). A hundred years of prime numbers, *Amer. Math. Monthly* **103**, no. 9, pp. 729–741.

Berndt, B. C. (1975). Elementary evaluation of $\zeta(2n)$, *Math. Mag.* **48**, no. 3, pp. 148–154.

Bernstein, S. N. (1912a). Démonstration du théorème de Weierstrass fondée sur le calcul de probabilité, *Comm. Soc. Math. Kharkov* **13**, pp. 1–2.

———————— (1912b). Sur l'ordre de la meilleure approximation des fonctions continues par des polynomes de degré donné, *Acad. Roy. Belgique Cl. Sci. Mém. Coll. in-4°* (2) **4**, pp. 1–103.

———————— (1928). Sur les fonctions absolument monotones, *Acta Math.* **52**, pp. 1–66.

———————— (1931). Sur les polynômes orthogonaux relatifs à un segment fini II, *J. Math. Pures Appl.* (9) **10**, pp. 219–286.

Beukers, F. (1978). A note on the irrationality of $\zeta(2)$ and $\zeta(3)$, *Bull. London Math. Soc.* **11**, pp. 268–272.

Beukers, F., Calabi, E. and Kolk, J. A. C. (1993). Sums of generalized harmonic series and volumes, *Nieuw Arch. Wisk.* (4) **11**, pp. 217–224.

Binet, J. (1839). Mémoire sur les intégrales définies eulériennes, et sur leur application à la théorie des suites, ainsi qu'à l'évaluation des fonctions des grands nombres, *J. École Roy. Polytech.* **16**, pp. 123–343. See also C. R. Acad. Sci. Paris, **9** (1839), pp. 39–45.

Bohman, H. (1952). On approximation of continuous and of analytic functions, *Ark. Mat.* **2**, no. 3, pp. 43–56.

Bohr, H. and Mollerup, J. (1922). Lærebog i matematisk Analyse (A textbook of mathematical analysis), III, Grænseprocesser, Jul. Gjellerups Forlag, København.

Borel, E. (1895). Sur quelques points de la théorie des fonctions, *Ann. Sci. École Norm. Sup.* (3) **12**, pp. 9–55 (in particular, p. 44).

Boyd, D. W. (1969). Transcendental numbers with badly distributed powers, *Proc. Amer. Math. Soc.* **23**, no. 2, pp. 424–427.

Brown, G. and Koumandos, S. (1997). On a monotonic trigonometric sum, *Monatsh. Math.* **123**, pp. 109–119.

Caianiello, E. R. (1961). Outline of a theory of thought-processes and thinking machine, *J. Theoret. Biol.* **2**, pp. 204–235.

Callahan, F. P. (1964). Density and uniform density, *Proc. Amer. Math. Soc.* **15**, no. 5, pp. 841–843.

Carleman, T. (1922). Sur les fonctions quasi-analytiques, Proc. 5th Scand. Math. Congress, Helsingfors, Finland, pp. 181–196.

    = Édition complète des articles de Torsten Carleman, Litos Reprotryck, Malmö,

1960, pp. 199–214.

————— (1923). Über die Approximation analytischer Funktionen durch linear Aggregate von vorgegebenen Potenzen, *Ark. Mat. Astr. Fys.* **17**, no. 9, pp. 1–30.

————— (1927). Sur un théorème de Weierstrass, *ibid.* **20B**, no. 4, pp. 1–5.

Carleson, L. (1954). A proof of an inequality of Carleman, *Proc. Amer. Math. Soc.* **5**, no. 6, pp. 932–933.

Carlitz, L. (1961). A recurrence formula for $\zeta(2n)$, *Proc. Amer. Math. Soc.* **12**, no. 6, pp. 991–992.

Carlson, F. (1935). Une inégalité, *Ark. Mat. Astr. Fys.* **25B**, no. 1, pp. 1–5.

Catalan, E. (1875). Sur la constante d'Euler et la fonction de Binet, *J. Math. Pures Appl.* (3) **1**, pp. 209–240.

Cesàro, E. (1888). *Nouv. Ann. Math.* (3) **17**, p. 112.

————— (1893). Sulla determinazione assintotica delle serie di potenze, *Atti R. Accad. Sc. Fis. Mat. Napoli* (2) **7**, pp. 187–195.

= Ernesto Cesàro, Opere Scelte, Vol. 1, Parte seconda, Ed. Cremonese, Roma, 1965, pp. 397–406.

————— (1906). Fonctions continues sans dérivée, *Arch. Math. Phys.* (3) **10**, pp. 57–63.

Chebyshev, P. L. (1852). Mémoire sur les nombres premiers, Œuvres de P. L. Tchebychef, Tome I, Chelsea Pub. Co., New York, 1961, pp. 51–70.

————— (1854). Théorie des mécanismes connus sous le nom de parallélogrammes, Œuvres de P. L. Tchebychef, Tome I, Chelsea Pub. Co., New York, 1961, pp. 111–143.

————— (1859). Sur l'interpolation dans le cas d'un grand nombre de données fournies par les observations, Œuvres de P. L. Tchebychef, Tome I, Chelsea Pub. Co., New York, 1961, pp. 387–469.

————— (1881). Sur les fonctions qui s'écartent peu de zéro pour certaines valeurs de la variable, Œuvres de P. L. Tchebychef, Tome II, Chelsea Pub. Co., New York, 1961, pp. 335–356.

Chen, M.-P. (1975). An elementary evaluation of $\zeta(2m)$, *Chinese J. Math.* **3**, no. 1, pp. 11–15.

Choe, B. R. (1987). An elementary proof of $\sum_{n=1}^{\infty} 1/n^2 = \pi^2/6$, *Amer. Math. Monthly* **94**, no. 7, pp. 662–663.

Choi, J. and Rathie, A. K. (1997). An evaluation of $\zeta(2)$, *Far East J. Math. Sci.* **5**, no. 3, pp. 393–398.

Choi, J., Rathie, A. K. and Srivastava, H. M. (1999). Some hypergeometric and other evaluations of $\zeta(2)$ and allied series, *Appl. Math. Comput.* **104**, no. 2-3, pp. 101–108.

Clausen, T. (1840). Lehrsatz aus einer Abhandlung über die Bernoullischen Zahlen, *Astronomische Nachrichten* **17**, no. 406, pp. 351–352.

Davis, P. J. (1959). Leonhard Euler's integral : a historical profile of the Gamma function, *Amer. Math. Monthly* **66**, no. 10, pp. 849–869.

de Boor, C. and Schoenberg, I. J. (1976). Cardinal interpolation and spline functions VIII. The Budan-Fourier theorem for splines and applications, Lect. Notes in Math., 501, pp. 1–77.

de la Vallée Poussin, Ch.-J. (1896). Recherches analytiques sur la théorie des nombres premiers. Première partie: La fonction $\zeta(s)$ de Riemann et les nombres premiers en général, suivi d'un Appendice sur des réflexions applicables à une formule donnée par Riemann, *Ann. Soc. Sci. Bruxelles* (deuxième partie) **20**, pp. 183–256.

= Charles-Jean de La Vallée Poussin Collected Works, Vol. I, Acad. Roy. Belgique, Circolo Mat. Palermo, 2000, pp. 223–296.

de Rham, G. (1957). Sur un exemple de fonction continue sans dérivée, *Enseign. Math.* (2) **3**, pp. 71–72.

Denquin, C. (1912). Sur quelques séries numériques, *Nouv. Ann. Math.* (4) **12**, pp. 127–135.

Diamond, H. G. (1982). Elementary methods in the study of the distribution of prime numbers, *Bull. Amer. Math. Soc.* (N.S.) **7**, no. 3, pp. 553–589.

Duncan, J. and McGregor, C. M. (2003). Carleman's inequality, *Amer. Math. Monthly* **110**, no. 5, pp. 424–431.

Duffin, R. J. and Schaeffer, A. C. (1941). A refinement of an inequality of the brothers Markoff, *Trans. Amer. Math. Soc.* **50**, no. 3, pp. 517–528.

Elkies, N. D. (2003). On the sums $\sum_{k=-\infty}^{\infty}(4k + 1)^{-n}$, *Amer. Math. Monthly* **110**, no. 7, pp. 561–573.

Erdös, T. (1949). On a new method in elementary number theory which leads to an elementary proof of the prime number theorem, *Proc. Nat. Acad. Sci. USA* **35**, pp. 374–384.

Estermann, T. (1947). Elementary evaluation of $\zeta(2k)$, *J. London Math. Soc.* **22**, pp. 10–13.

Faber, G. (1907). Einfaches Beispiel einer stetigen nirgends differentiierbaren Funktion, *Jber. Deutsch. Math. Verein.* **16**, pp. 538–540.

Fejér, L. (1900). Sur les fonctions bornées et intégrables, *C. R. Acad. Sci. Paris* **131**, pp. 984–987.

———— (1910). Über gewisse Potenzreihen an der Konvergenzgrenze, *S.-B. math.-phys. Akad. Wiss. München* **40**, no. 3, pp. 1–17.

= Leopold Fejér Gesammelte Arbeiten, Band I, Birkhäuser, 1970, pp. 573–583.

———— (1925). Abschätzungen für die Legendreschen und verwandte Polynome, *Math. Z.* **24**, pp. 285–298.

Fekete, M. (1923). Über die Verteilung der Wurzeln bei gewissen algebraischen Gleichungen mit ganzzahligen Koeffizienten, *Math. Z.* **17**, pp. 228–249 (in particular, p. 233).

Franklin, F. (1885). Proof of a theorem of Tchebycheff's on definite integrals, *Amer. J. Math.* **7**, no. 4, pp. 377–379.

Gibbs, J. W. (1899). letter to the editor, Fourier Series, *Nature* **59**, p. 200 and p. 606.

Giesy, D. P. (1972). Still another elementary proof that $\sum 1/k^2 = \pi^2/6$, *Math. Mag.* **45**, no. 3, pp. 148–149.

Goldscheider, F. (1913). *Arch. Math. Phys.* **20**, pp. 323–324.

Goldstein, L. J. (1973). A history of the prime number theorem, *Amer. Math. Monthly* **80**, no. 6, pp. 599–615.

Gronwall, T. H. (1912). Über die Gibbssche Erscheinung und die trigonometrischen Summen $\sin x + \dfrac{1}{2}\sin 2x + \cdots + \dfrac{1}{n}\sin nx$, *Math. Ann.* **72**, pp. 228–243.

———————— (1913). Über die Laplacesche Reihe, *ibid.* **74**, pp. 213–270.

———————— (1918). The gamma function in the integral calculus, *Ann. Math.* (2) **20**, no. 2, pp. 35–124.

Grosswald, E. (1980). An unpublished manuscript of Hans Rademacher, *Historia Math.* **7**, no. 4, pp. 445–446.

Hadamard, J. (1896). Sur la distribution des zéros de la fonction $\zeta(s)$ et ses conséquences arithmétiques, *Bull. Soc. Math. France* **24**, pp. 199–220.

———— (1914). Sur le module maximum d'une fonction et de ses dérivées, *ibid.* **42**, pp. 68–72.

= Œuvres de Jacques Hadamard, Tome I, Edition du C. N. R. S., Paris, 1968, pp. 379–382.

Hardy, G. H. (1912). Note on Dr. Vacca's series for γ, *Quart. J. Math.* **43**, pp. 215–216.

———— (1914). Sur les zéros de la fonction $\zeta(s)$ de Riemann, *C. R. Acad. Sci. Paris* **158**, pp. 1012–1014.

Hardy, G. H. and Littlewood, J. E. (1914). Tauberian theorems concerning power series and Dirichlet's series whose coefficients are positive, *Proc. London Math. Soc.* (2) **13**, pp. 174–191.

———————————————— (1914). Some problems of Diophantine approximation, *Acta Math.* **37**, pp. 155–190.

Harper, J. D. (2003). Another simple proof of $1 + 1/2^2 + 1/3^2 + \cdots = \pi^2/6$, *Amer. Math. Monthly* **110**, no. 6, pp. 540–541.

Hata, M. (1982). Dynamics of Caianiello's equation, *J. Math. Kyoto Univ.* **22**, no. 1, pp. 155–173.

———— (1995). Farey fractions and sums over coprime pairs, *Acta Arith.* **70**, no. 2, pp. 149–159.

Hausdorff, F. (1925). Zum Hölderschen Satz über $\Gamma(x)$, *Math. Ann.* **94**, pp. 244–247.

Hecke, E. (1921). Über analytische Funktionen und die Verteilung von Zahlen mod. eins, *Abh. Math. Sem. Hamburg Univ.* **1**, pp. 54–76.

= Erich Hecke Mathematische Werke, Göttingen, 1959, pp. 313–335.

Hofbauer, J. (2002). A simple proof of $1 + 2^{-2} + 3^{-2} + \cdots = \pi^2/6$ and related identities, *Amer. Math. Monthly* **109**, no. 2, pp. 196–200.

Hölder, O. (1887). Ueber die Eigenschaft der Gammafunction keiner algebraischen Differentialgleichung zu genügen, *Math. Ann.* **28**, pp. 1–13.

Holme, F. (1970). En enkel beregning av $\sum_{k=1}^{\infty} 1/k^2$, *Nordisk Mat. Tidskr.* **18**, pp. 91–92.

Hovstad, R. M. (1972). The series $\sum_{k=1}^{\infty} 1/k^{2p}$, the area of the unit circle and Leibniz' formula, *Nordisk Mat. Tidskr.* **20**, pp. 92–98.

Hua, L.-G. (1965). On an inequality of Opial, *Sci. Sinica* **14**, pp. 789–790.

Jackson, D. (1911). Über eine trigonometrische Summe, *Rend. Circ. Math. Palermo* **32**, pp. 257–262.

Jensen, J. L. W. V. (1906). Sur les fonctions convexes et les inégalités entre les values moyennes, *Acta Math.* **30**, pp. 175–193.

———————————————— (1916). An elementary exposition of the theory of the Gamma function. Authorized translation from the Danish by T. H. Gronwall, *Ann. Math.* (2) **17**, no. 3, pp. 124–166.

Ji, C.-G. and Chen, Y.-G. (2000). Euler's formula for $\zeta(2k)$, proved by induction on k, *Math. Mag.* **73**, no. 2, pp. 154–155.

Kakeya, S. (1914). On approximate polynomials, *Tôhoku Math. J.* **6**, pp. 182–186.

Kalman, D. (1993). Six ways to sum a series, *College Math. J.* **24**, no. 5, pp. 402–421.

Karamata, J. (1930). Über die Hardy-Littlewoodschen Umkehrungen des Abelschen

Stätigkeits-satzes, *Math. Z.* **32**, pp. 319–320.

Katznelson, Y. and Stromberg, K. (1974). Everywhere differentiable, nowhere monotone, functions, *Amer. Math. Monthly* **81**, no. 4, pp. 349–354.

Kempner, A. J. (1914). A curious convergent series, *Amer. Math. Monthly* **21**, no. 2, pp. 48–50.

Khintchine, A. (1924). Über einen Satz der Wahrscheinlichkeitsrechnung, *Fund. Math.* **6**, pp. 9–20.

Kimble, G. (1987). Euler's other proof, *Math. Mag.* **60**, no. 5, p. 282.

Klamkin, M. S. (1970). A comparison of integrals, *Amer. Math. Monthly* **77**, no. 10, p. 1114; ibid. **78**, no. 6, pp. 675–676.

Kluyver, J. C. (1900). Der Staudt-Clausen'sche Satz, *Math. Ann.* **53**, pp. 591–592.

————— (1928). Über Reihen mit positiven Gliedern, *J. London Math. Soc.* **3**, pp. 205–211.

Knopp, K. and Schur, I. (1918). Über die Herleitung der Gleichung $\sum_{n=1}^{\infty} 1/n^2 = \pi^2/6$, *Arch. Math. Phys.* (3) **27**, pp. 174–176.

Kolmogorov, A. N. (1939). On inequalities for suprema of consecutive derivatives of an arbitrary function on an infinite interval (Russian), *Uchenye Zapiski Moskov. G. Univ. Mat.* **30**, no. 3, pp. 3–13.

= Amer. Math. Soc. Transl., Ser. 1, Vol. 2, 1962, pp. 233–243.

= Selected Works of A. N. Kolmogorov, Vol. 1, ed. V. Tikhomirov, Kluwer, 1991, pp. 277–290.

Köpcke, A. (1887). Ueber Differentiirbarkeit und Anschaulichkeit der stetigen Functionen, *Math. Ann.* **29**, pp. 123–140.

————— (1889). Ueber eine durchaus differentiirbare, stetige Function mit Oscillationen in jedem Intervalle, *ibid.* **34**, pp. 161–171.

————— (1890). Nachtrag zu dem Aufsatze „Ueber eine durchaus differentiirbare, stetige Function mit Oscillationen in jedem Intervalle" (Annalen, Band XXXIV, pag. 161 ff.), *ibid.* **35**, pp. 104–109.

Korevaar, J. (1982). On Newman's quick way to the prime number theorem, *Math. Intelligencer* **4**, no. 3, pp. 108–115.

Korkin, A. N. and Zolotareff, E. I. (1873). Sur une certain minimum, *Nouv. Ann. Math.* (2) **12**, pp. 337–355.

Korovkin, P. P. (1953). On convergence of linear positive operators in the space of continuous functions (Russian), *Dokl. Akad. Nauk SSSR* **90**, no. 6, pp. 961–964.

Kortram, R. A. (1996). Simple proofs for $\sum_{k=1}^{\infty} 1/k^2 = \pi^2/6$ and $\sin x = x \prod_{k=1}^{\infty}(1 - x^2/(k\pi)^2)$, *Math. Mag.* **69**, no. 2, pp. 122–125.

Koumandos, S. (2001). Some inequalities for cosine sums, *Math. Inequal. Appl.* **4**, no. 2, pp. 267–279.

Kummer, E. E. (1847). Beitrag zur Theorie der Function $\Gamma(x) = \int_0^{\infty} e^{-v} v^{x-1}\, dv$, *J. Reine Angew. Math.* **35**, pp. 1–4.

Kuo, H.-T. (1949). A recurrence formula for $\zeta(2n)$, *Bull. Amer. Math. Soc.* **55**, pp. 573–574.

Landau, E. (1908). Über die Approximation einer stetigen Funktionen durch eine ganze rationale Funktion, *Rend. Circ. Mat. Palermo* **25**, pp. 337–345.

————— (1913). Einige Ungleichungen für zweimal differentiierbare Funktionen, *Proc.*

London Math. Soc. (2) **13**, pp. 43–49.

————— (1915). Über die Hardysche Entdeckung unendlich vieler Nullstellen der Zeta-funktion mit reellem Teil $\frac{1}{2}$, *Math. Ann.* **76**, pp. 212–243.

————— (1934). Über eine trigonometrische Ungleichung, *Math. Z.* **37**, p. 36.

Landsberg, G. (1908). Über Differentiierbarkeit stetiger Funktionen, *Jber. Deutsch. Math. Verein.* **17**, pp. 46–51.

Lerch, M. (1888). Ueber die Nichtdifferentiirbarkeit gewisser Functionen, *J. Reine Angew. Math.* **103**, pp. 126–138.

————— (1903). Sur un point de la théorie des fonctions génératrices d'Abel, *Acta Math.* **27**, pp. 339–352.

Levinson, N. (1966). On the elementary proof of the prime number theorem, *Proc. Edinburgh Math, Soc.* (2) **15**, pp. 141–146.

————— (1969). A motivated account of an elementary proof of the prime number theorem, *Amer. Math. Monthly* **76**, no. 3, pp. 225–245.

Lebesgue, H. (1898). Sur l'approximation des fonctions, *Bull. Sci. Math.* **22**, pp. 278–287.

Littlewood, J. E. (1910). The converse of Abel's theorem on power series, *Proc. London Math. Soc.* (2) **9**, pp. 434–448.

Macdonald, I. G. and Nelsen, R. B. (1979). E2701, *Amer. Math. Monthly* **86**, no. 5, p. 396.

Malmstén, C. J. (1847). Sur la formule $hu'_x = \Delta u_x - \frac{h}{2}\Delta u'_x + \frac{B_1 h^2}{1.2} \cdot \Delta u''_x - \frac{B_2 h^4}{1...4}\Delta u_x^{IV} +$ etc, *J. Reine Angew. Math.* **35**, no. 1, pp. 55–82.

Markov, A. A. (1889). On a problem of D. I. Mendeleev (Russian), *Zap. Imp. Akad. Nauk, St. Petersburg* **62**, pp. 1–24.

Markov, V. A. (1892). On functions deviating the least from zero on a given interval (Russian), Depart. Appl. Math. Imperial St. Petersburg Univ.
= Über Polynome, die in einem gegebenen Intervalle möglichst wenig von Null abweichen, *Math. Ann.* **77**, pp. 213–258, 1916.

Matsuoka, Y. (1961). An elementary proof of the formula $\sum_{k=1}^{\infty} 1/k^2 = \pi^2/6$, *Amer. Math. Monthly* **68**, no. 5, pp. 485–487.

Mehler, F. G. (1872). Notiz über die Dirichlet'schen Integralausdrücke für die Kugelfunction $P^n(\cos\vartheta)$ und über eine analoge Integralform für die Cylinderfunction $J(x)$, *Math. Ann.* **5**, pp. 141–144.

Mertens, F. (1875). Ueber die Multiplicationsregel für zwei unendliche Reihen, *J. Reine Angew. Math.* **79**, pp. 182–184.

Michelson, A. A. (1898). letter to the editor, Fourier Series, *Nature* **58**, pp. 544–545.

Mirkil, H. (1956). Differentiable functions, formal power series, and moments, *Proc. Amer. Math. Soc.* **7**, no. 4, pp. 650–652.

Mittag-Leffler, G. (1900). Sur la représentation analytique des fonctions d'une variable réelle, *Rend. Circ. Mat. Palermo* **14**, pp. 217–224.

Möbius, A. F. (1832). Über eine besondere Art von Umkehrung der Reihen, *J. Reine Angew. Math.* **9**, no. 2, pp. 105–123.

Moore, E. H. (1897). Concerning transcendentally transcendental functions, *Math. Ann.* **48**, pp. 49–74.

Müntz, Ch. H. (1914). Über den Approximationssatz von Weierstraß, Mathematische Abhandlungen Hermann Amandus Schwarz zu seinem fünfzigjährigen Doktorjubiläum am 6 August 1914 gewidmet von Freunden und Schülern, Berlin, J. Springer, pp. 303

−312 [There is a textually unaltered reprint by Chelsea Pub. Co. in 1974].

Neville, E. H. (1951). A trigonometrical inequality, *Proc. Cambridge Philos. Soc.* **47**, pp. 629−632.

Newman, D. J. (1980). Simple analytic proof of the prime number theorem, *Amer. Math. Monthly* **87**, no. 9, pp. 693−696.

Nikolić, A. (2002). Jovan Karamata (1902−1967), *Novi Sad J. Math.* (1) **32**, pp. 1−5.

Niven, I. (1947). A simple proof that π is irrational, *Bull. Amer. Math. Soc.* **53**, p. 509.

Okada, Y. (1923). On approximate polynomials with integral coefficients only, *Tôhoku Math. J.* **23**, pp. 26−35.

Opial, Z. (1960). Sur une inégalité, *Ann. Polon. Math.* **8**, pp. 29−32.

Osler, T. J. (2004). Finding $\zeta(2p)$ from a product of sines, *Amer. Math. Monthly* **111**, no. 1, pp. 52−54.

Ostrowski, A. (1919). Neuer Beweis des Hölderschen Satzes, daß die Gammafunktion keiner algebraischen Differentialgleichung genügt, *Math. Ann.* **79**, pp. 286−288.

————— (1925). Zum Hölderschen Satz über $\Gamma(x)$, *ibid.* **94**, pp. 248−251.

Pál, J. (1914). Zwei kleine Bemerkungen, *Tôhoku Math. J.* **6**, pp. 42−43.

Papadimitriou, I. (1973). A simple proof of the formula $\sum_{k=1}^{\infty} k^{-2} = \pi^2/6$, *Amer. Math. Monthly* **80**, no. 4, pp. 424−425.

Pereno, I. (1897). Sulle funzioni derivabili in ogni punto ed infinitamente oscillati in ogni intervallo, *Giorn. di Mat.* **35**, pp. 132−149.

Picard, É. (1891). Sur la représentation approchée des fonctions, *C. R. Acad. Sci. Paris* **112**, pp. 183−186.

Pisot, Ch. (1938). La répartition modulo 1 et les nombres algébriques, *Ann. Sc. Norm. Sup. Pisa* (2) **7**, pp. 205−248.

————— (1946). Répartition (mod 1) des puissances successives des nombres réels, *Comment. Math. Helv.* **19**, pp. 153−160.

Pólya, G. (1911). *Nouv. Ann. Math.* (4) **11**, pp. 377−381.

————— (1926). Proof of an inequality, *Proc. London Math. Soc.* (2) **24**, p. vi in 'Records of Proceedings at Meetings'.

————— (1931). *Jber. Deutsch. Math. Verein.* **40**, p. 81.

Pringsheim, A. (1900). Ueber das Verhalten von Potenzreichen auf dem Convergenzkreise, *S.-B. math.-phys. Akad. Wiss. München* **30**, pp. 37−100.

Raabe, J. L. (1843). Angenäherte Bestimmung der Factorenfolge 1.2.3.4.5...$n = \Gamma(1+n) = \int x^n e^{-x} dx$, wenn n eine sehr grosse Zahl ist, *J. Reine Angew. Math.* **25**, pp. 146−159.

————— (1844). Angenäherte Bestimmung der Function $\Gamma(1+n) = \int_0^{\infty} x^n e^{-x} dx$, wenn n eine ganze, gebrochene, order incommensurable sehr grosse positive Zahl ist, *ibid.* **28**, pp. 10−18.

Rademacher, H. (1922). Einige Sätze über Reihen von allgemeinen Orthogonalfunktionen, *Math. Ann.* **87**, pp. 112−138.

Redheffer, R. (1967). Recurrent inequalities, *Proc. London Math. Soc.* (3) **17**, pp. 683−699.

Riemann, G. F. B. (1859). Ueber die Anzahl der Primzahlen unter einer gegebenen Grösse, *Bernhard Riemann's Gesammelte Mathematische Werke und Wissenschaftlicher Nachlass*, Zweite Auflage, Leipzig, 1892, pp. 145−153.

[English Translation: Collected Papers Bernhard Riemann, Kendrick Press, 2004, pp. 135–143.]

Robbins, N. (1999). Revisiting an old favorite: $\zeta(2m)$, *Math. Mag.* **72**, no. 4, pp. 317–319.

Rogosinski, W. (1955). Some elementary inequalities for polynomials, *Math. Gazette* **39**, pp. 7–12.

Rogosinski, W. and Szegö, G. (1928). Über die Abschnitte von Potenzreihen, die in einem Kreise beschränkt bleiben, *Math. Z.* **28**, pp. 73–94.

Rosenthal, A. (1953). On functions with infinitely many derivatives, *Proc. Amer. Math. Soc.* **4**, no. 4, pp. 600–602.

Runge, C. (1885a). Zur Theorie der eindeutigen analytischen Functionen, *Acta Math.* **6**, pp. 229–245.

———— (1885b). Über die Darstellung willkürlicher Funktionen, *ibid.* **7**, pp. 387–392.

Russell, D. C. (1991). Another Eulerian-type proof, *Math. Mag.* **64**, no. 5, p. 349.

Schlömilch, O. (1844). Einiges über die Eulerschen Integrale der zweiten Art, *Arch. Math. Phys.* (2) **4**, pp. 167–174.

Schönbeck, J. (2004). Thomas Clausen und die quadrierbaren Kreisbogenzweiecke, *Centaurus* **46**, pp. 208–229.

Schwering, K. (1899). Zur Theorie der Bernoulli'schen Zahlen, *Math. Ann.* **52**, pp. 171–173.

Selberg, A. (1949). An elementary proof of the prime-number theorem, *Ann. Math.* **50**, no. 2, pp. 305–313.

Skau, I. and Selmer, E. S. (1971). Noen anvendelser av Finn Holmes methode for beregning av $\sum_{k=1}^{\infty} 1/k^2$, *Nordisk Mat. Tidskr.* **19**, pp. 120–124.

Stäckel, P. (1913). *Arch. Math. Phys.* (3) **13**, p. 362.

Stark, E. L. (1969). Another proof of the formula $\sum 1/k^2 = \pi^2/6$, *Amer. Math. Monthly* **76**, no. 5, pp. 552–553.

———— (1972). A new method of evaluating the sums of $\sum_{k=1}^{\infty}(-1)^{k+1}k^{-2p}$, $p = 1, 2, 3, \ldots$ and related series, *Elem. Math.* **27**, no. 2, pp. 32–34.

———— (1974). The series $\sum_{k=1}^{\infty} k^{-s}$, $s = 2, 3, 4, \cdots$, once more, *Math. Mag.* **47**, no. 4, pp. 197–202.

Staudt, K. G. C. von (1840). Beweis eines Lehrsatzes, die Bernoullischen Zahlen betreffend, *J. Reine Angew. Math.* **21**, pp. 372–374.

Steinhaus, H. (1920). Sur les distances des points des ensembles de mesure positiv, *Fund. Math.* **1**, pp. 93–104.

Stieltjes, T. J. (1876). De la représentation approximative d'une fonction par une autre [traduction de la brochure imprimée à Delft en 1876], Œuvres complétes de Thomas Jan Stieltjes, Tome I, Noordhoff, Groningen, Netherlands, 1914, pp. 11–20.

———— (1890). Sur les polynômes de Legendre, Œuvres complétes de Thomas Jan Stieltjes, Tome II, Noordhoff, Groningen, Netherlands, 1914, pp. 236–252.

Szegö, G. (1934). *Jber. Deutsch. Math. Verein.* **43**, pp. 17–20.

———— (1948). On an inequality of P. Turán concerning Legendre polynomials, *Bull. Amer. Math. Soc.* **54**, pp. 401–405.

Takagi, T. (1903). A simple example of the continuous function without derivative, *Proc. Phys.-Math. Soc. Japan* (II) **1**, pp. 176–177.

Tatuzawa, T. and Iseki, K. (1951). On Selberg's elementary proof of the prime-number theorem, *Proc. Japan Acad.* **27**, pp. 340–342.

Tauber, A. (1897). Ein Satz aus der Theorie der unendlichen Reihen, *Monatsh. Math.* **8**, pp. 273–277.

Titchmarsh, E. C. (1926). A series inversion formula, *Proc. London Math. Soc.* (2) **26**, pp. 1–11.

Tsumura, H. (2004). An elementary proof of Euler's formula for $\zeta(2m)$, *Amer. Math. Monthly* **111**, no. 5, pp. 430–431.

Turán, P. (1950). On the zeros of the polynomials of Legendre, *Časopis Pěst. Mat. Fys.* **75**, pp. 113–122.

= Collected Papers of Paul Turán, Vol. 1, Edited by Paul Erdös, Akad. Kiadó, Budapest, 1990, pp. 531–540.

Underwood, R. S. (1928). An expression for the summation $\sum_{m=1}^{n} m^p$, *Amer. Math. Monthly* **35**, no. 8, pp. 424–428.

Vacca, G. (1910). A new series for the Eulerian constant, *Quart. J. Pure Appl. Math.* **41**, pp. 363–368.

van der Waerden, B. L. (1930). Ein einfaches Beispiel einer nicht-differenzierbaren stetigen Funktion, *Math. Z.* **32**, pp. 474–475.

Verblunsky, S. (1945). On positive polynomials, *J. London Math. Soc.* **20**, pp. 73–79.

Vijayaraghavan, T. (1940). On the fractional parts of the powers of a number (I), *J. London Math. Soc.* **15**, pp. 159–160.

———————— (1941). — (II), *Proc. Cambridge Philos. Soc.* **37**, pp. 349–357.

———————— (1942). — (III), *J. London Math. Soc.* **17**, pp. 137–138.

———————— (1948). — (IV), *J. Indian Math. Soc.* **12**, pp. 33–39.

Volterra, V. (1897). Sul principio de Dirichlet, *Rend. Circ. Mat. Palermo* **11**, pp. 83–86.

von Mangoldt, H. (1895). Zu Riemanns Abhandlung „Ueber die Anzahl der Primzahlen unter einer gegebenen Grösse", *J. Reine Angew. Math.* **114**, no. 3-4, pp. 255–305.

Walsh, J. L. (1923). A closed set of normal, orthogonal functions, *Amer. J. Math.* **45**, no. 1, pp. 5–24.

Weierstrass, K. (1856). Über die Theorie der analytischen Facultäten, *J. Reine Angew. Math.* **51**, pp. 1–60.

———————— (1885). Über die analytische Darstellbarkeit sogenannter willkürlicher Funktionen reeller Argumente, *S.-B. Königl. Akad. Wiss. Berlin,* pp. 633–639 and pp. 789–805.

= Mathematische Werke von Karl Weierstrass, Band III, Berlin, Mayer & Müller, 1903, pp. 1–37.

Weyl, H. (1916). Über die Gleichverteilung von Zahlen mod. Eins, *Math. Ann.* **77**, pp. 313–352.

Wielandt, H. (1952). Zur Umkehrung des Abelschen Stetigkeitssatzes, *Math. Z.* **56**, no. 2, pp. 206–207.

William Lowell Putnam Mathematical Competition (1970). A-4, *Amer. Math. Monthly* **77**, no. 7, p. 723 and p. 725.

Williams, G. T. (1953). A new method of evaluating $\zeta(2n)$, *Amer. Math. Monthly* **60**, no. 1, pp. 19–25.

Williams, K. S. (1971). On $\sum_{n=1}^{\infty} (1/n^{2k})$, *Math. Mag.* **44**, no. 5, pp. 273–276.

Wright, E. M. (1952). The elementary proof of the prime number theorem, *Proc. Roy. Soc. Edinburgh Sect. A* **63**, pp. 257–267.

———————— (1954). An inequality for convex functions, *Amer. Math. Monthly* **61**, no. 9, pp. 620–622.

Yaglom, A. M. and Yaglom, I. M. (1953). An elementary derivation of the formula of
Wallis, Leibniz and Euler for the number π (Russian), *Uspehi Mat. Nauk* (N.S.) **8**,
no. 5 (57), pp. 181–187.

Young, W. H. (1909). On differentials, *Proc. London Math. Soc.* (2) **7**, pp. 157–180.

——————— (1912). On a certain series of Fourier, *ibid.* (2) **11**, pp. 357–366.

Zagier, D. (1997). Newman's short proof of the prime number theorem, *Amer. Math.
Monthly* **11**, no. 8, pp. 705–708.

Index

Bibliography   291